"The Origin of Man's Ethical Behavior"
(October 1941 title)

"Ethics and the Struggle for Existence"
(April 1941 title)

Ernest Everett Just and Hedwig Schnetzler Just

This archival edition is 251 pages (including an addendum with 8 other pages) from a missing since 1941 unpublished book manuscript "The Origin of Man's Ethical Behavior" co-authored by Ernest Everett Just and Hedwig Anna Schnetzler Just, discovered in 2018 preserved among the collected papers of E. E. Just at the Howard University Moorland-Spingarn Research Center, compiled, transcribed, and gently edited in 2018, 2019, and 2020 by theological ethicist Theodore Walker Jr. and archival researcher Lillie R. Jenkins, with additional co-editing by Walker, Jenkins, and biochemist and E. E. Just scholar W. Malcolm Byrnes; in consultation with cell biologist and E. E. Just scholar Stuart Newman, historian of science and E. E. Just-scholar-biographer Kenneth R. Manning, historian of religions Charles H. Long, and Moorland-Spingarn curator of manuscripts Joellen ElBashir.

3

Page 1
[Box 125-19, folder 382, contains most of pp. 1-251 (with pp. 1-203 being origin manuscript page numbers).]

Chapter 1 - **THE PROBLEM STATED** – pp. 1-17

Many of us include in the term, ethics, convention and custom and consider as ethical laws those established by a set of <u>mores</u> whence are derived moral principles. These <u>mores</u> may differ in different lands to such an extent that conduct regarded as ethical or moral in one is a serious infraction of decency in another. Moreover, in a given country customs change in the course of history: what once was interdicted becomes socially correct. For manners, like language, change, often in favor of what at first was frowned upon as inelegant.

Ethics is sometimes further limited in being used by a special class and thus carries a limited meaning. When the physician or surgeon speaks of his professional ethics, he refers to certain rules of conduct within the profession, infractions of which are held to be serious. Etiquette in other instances is confused with ethics. Morality is often used as a synonym of ethics. Morality, or better immorality, in America generally, refers to the relation between the sexes.

Now ethics broadly considered as belonging either to the domain of philosophy or of religion or of a territory common to both has a meaning wider than geographical and political boundaries. It cannot be local as to time or to place. It is a perennial problem of universality common to all mankind, savage and civilized alike. At the same time it transcends philosophy as nowadays of restricted meaning despite the fact that philosophy of the older days including all knowledge has given us most abundant ethical concepts, concepts which have varied both as to time and as to place in the history of philosophy.

Philosophy as the codifying of all knowledge endeavors to reach truth largely through thinking and by means of logic. It takes as raw material the experiences of mankind which it reduces to generalizations. Systems of ethics devised by philosophers in the main show a lack of sufficient evidence gathered by natural science and tend to be based too much on ratiocination. This is all the more true the greater the degree to which philosophy has lost its original domain to what have become separate sciences, and as philosophy has become more restricted to metaphysics. Thus we do not wish to state our problem in terms of one or another philosopher. This is the case not only because there is no unanimity among philosophers on the subject; but also because we intend to formulate the problem in, if you please, a non-philosophical, i.e., non-metaphysical, sense. That is, we intend to treat ethics as a problem in biology, in physiology even.

This being the intention, we dismiss likewise all propositions concerning the problem of ethics which postulate it as being in the realm of any religion. It is of no moment to us that a popular idea still persists that ethics is a subject

Page 3

in religion and that only priests or old men of the tribe are teachers of it. True, founders of religion have often been great teachers of ethics. Christ was one of the greatest of these. Ethics is not religion therefore—any more than the Christian church as an institution, or all modern Christianity, for that matter, represents the religious teachings of Christ.

Religion is far too narrow a base upon which to superpose the structure of ethics. One religion or all religions taken together do not include all mankind. Furthermore, religion is a repository of aspirations, yearnings, common to its members. It makes a common fund of these deeper feelings which is easily accessible to all the devotees. It recognizes and stimulates what is already present in man's soul. Itself, it creates nothing.

It is within the field of biology, then, that we locate human ethics, or better to say, man's ethical behavior. The latter term is preferred for the following reasons.

In the first place, the term, ethics, however exactly defined retains hazy connotations. The exposition in the following pages attempts neither to elaborate a new system of ethics nor to derive a new meaning for the term. Nevertheless, because various schools of ethics have in formulating their concepts so emphasized each its individual point of view

Page 4

that the reader of them easily thinks of ethics as of several species instead of the general features (of common to) the problem, it becomes necessary in a discussion involving these general features of the problem, and only them, to substitute for the term ethics the less fleeting and more definite term, man's ethical behavior.

Secondly, the term, ethical behavior, connotes more sharply the manifestation of an activity. Not with ethics as body of laws or as science are we primarily concerned. Rather, we examine reactions as processes of responses to an imponderable. Comparably, the physicist deals with electricity or with force. More closely, the biologist should make, though rarely he does, the distinction between embryology, Lehre, and embryogenesis, Vorgang. Given this imponderable, the question is, how does man, a biological unit, respond to it, that is, what is his behavior? We search for the complex in man which vibrates in harmony to something outside of himself, something which is of himself as is the raw stuff, food, that makes him. Because ethics can be ascribed to the domain of biological phenomena, of vital manifestations, the term, behavior, is admissible. However, the term as used here carries no implication of behavior psychology.

Finally, the term, ethical behavior, is preferable to the term, ethics, because we purpose to examine the origin of that upon which the body, principle or abstract concept, ethics, is derived. Human behavior involves the whole vital unit. It is thus never divorced from the human spirit.

Page 5

In matters of the spirit, the real man, the thinking man brooks no interference. He defies spiritual domination; he denies authoritarianism. In his whole search for truth in the infinitely small, in the endlessness of space, encompassing other worlds than this, man enjoys his only freedom—that, which is his only real possession, the freedom of the soul. In this illimitable domain, every man is both monarch and creator. When as in all history attempts are made to dispossess him of this realm of his own creation, these all fail. Such attempts give birth to a fiercer allegiance, a more burning passion to exercise the right of domain. The temporary threats to soul-freedom serve merely to accentuate the diapason yet to come, which must come, in greater volume, in fuller tone. To man alone in this thinking is the whole universe in all time a dwelling-place. This spirit is no sudden endowment. Rather, it represents at its highest the whole history of the universe condensed into a single expression. From dust to dust again is the fate of man's corporeal being—his spirit is the music of the ages that never dies but rises again waveringly to the stars to become again one with the essence of that whence it came. It is to this ever-expanding, never to be checked overflowing of man's spirit that we relate the origin of man's ethical behavior; in that whence came his manifold interdependence with all nature, we seek its origin.

Page 6

Now certainly, no one pretends that he can prove beyond question the origin of man's ethical behavior. The origin of man's corporeal being is itself obscured in the mists of the unknowable. No experiment has yet been devised that can provide a basis for deriving the proof of the origin of man as substance. There is less likelihood that experience will reveal to us the beginning of his spiritual being. Thus, we can only continue today the argument and the speculations of the past concerning the origin of man's ethical behavior.

But some of us make votive offerings to exact science, and subscribe blindly, with an all-abiding faith, to the fixity and to the finality of scientific concepts, and accept as perfectly proved the inalterability of natural laws, paying vain oblations to the god of "scientific method" as omniscient and omnipotent. These devout ones are found not only among the non-scientific but, unfortunately, among professed scientists as well who by exaggerating scientific achievements render incalculable harm both to science and to mankind. To such as these an essay on so intangible a problem as the origin of ethical behavior is a lamentable waste of time. Even those who do not make extravagant claims for the all-power of science may ask, what gain is there in speaking of ethics at all? To them, as to most of us, this is the age of science, the triumph of materialism. But despite the glamorous exactitude of modern natural science which we admit and admire, natural science is not all of science. And all the sciences together do not comprise all of life. Science, however exact, ought not to create for us a realm wherein we wander as sleep-walkers in a dream. Material science procreates only more material science, more exact, more refined; more lamps and better lamps, more telescopes and bigger ones; greater nicety in the measurements of the electron. It will doubtless continue its advancement to the end of time. And it is the mind that makes possible this scientific pro-creation. It thus appears to be erroneous to set off exactitude in science, especially in natural science, and make it all of life.

Lamarck once said that man has two interests, the economical and the philosophical. These interests science satisfies. It is therefore a mistake to see in science the ends which are its means. Exact and pure science aims finally to satisfy our two interests. Hence, we have the right to speculate on problems that are incapable of direct measurement and of experiment. We do not thereby hypothecate science, unless for us, science has too restricted a meaning, in addition to our picture of it wherein we confuse means and ends. The biologist who accepts a too limited definition of science that is made in terms of exact measurements must perforce abandon his calling, for he dismisses the very basis of his science, the theory of evolution. Surely, in terms of modern physics, the ultra-experimental science, the theory of evolution is incapable of proof. But let me dismiss the case as presented by those who defend a brief for modern science with which I cannot agree. Let me take instead the position held by scientists of more modesty.

Page 8

Few physical scientists have the temerity to assert that what they call laws in physics are final and unalterable; they do not assume that they can prove these laws beyond question. In biology the possibilities of absolute proof are far less. Science may in large measure be defined as the knowledge of relationships. This knowledge is surer the more quantitative these relationships may be expressed. Physics thus is the most knowable, chemistry next. Biology is far from being capable of being expressed in quantitative terms. The conception of science as comprising data absolutely proved is not that of scientists themselves worthy of the name. We can prove nothing really.

All of this, let it be understood, does not place science under an interdict. To the contrary. From science, the knowable, depend our speculations concerning the unknowable. Indulgence in tergiversation by travel, along the way of inknowable to unknowable, seems to me a more hazardous journey. Rather, as in most exact science, by extrapolation, one extends knowledge into a new region of the material, we may move out from the realm of the concretely materialistic into that of the spiritual. For me, spiritual (social) science is but an extrapolation of the natural—with the difference that the variables are greater in number. Man as spiritual being we know only as a manifestation of his material being. However real we conceive his spirituality, as the only reality even, it remains intangible and refractory to analysis except as a behavior of man, the animal, the <u>zoological</u> entity and the biological even.

From this point of view, whence we envisage our problem, we cannot invoke the accepted criteria of exact natural science, and these failing, dismiss the case for the origin of man's ethical behavior. Despite ignorance, in the sense of the lack of proof—even if proof be used in the loosest sense—of the origin of man's corporeal being, we can nonetheless state a plausible case for it. This is built on the assumption that the world of living things has evolved. Not only do we locate our problem within the framework of biology, but also we base it on the theory of evolution.

In the following pages, I aim to establish a thesis of the origin of man's ethical behavior on the basis of the assumption that the world of living things is the product of evolution. Thus, at the outset I dismiss the theory of a special creation of each and every living thing that has existed, now exists, and will ever exist. This total rejection of the theory of special creation is fundamental for my thesis. Not to be clear at the outset concerning the point of departure is to vitiate the whole argument.

Here two points should be emphasized.

Although the proposition, that the world of living things is the product of evolution, cannot be proved, it is none the less capable of defense by a large body of evidence. The theory of special creation is less capable of such defense; it is instead an article of belief, i.e., of faith in a religious sense. We do not <u>know</u>, may never know, how living things came to be, whether through evolution or by special creation; yet we can accept one or the other proposition as the more reasonable.

The second point is that acceptance of the evidence in support of evolution does not entail atheism. Many devotees of religion admit the evidence for evolution without violation to belief in God. These beliefs are found in all camps of the multitudinous Christian cults, except those which insist upon a strictly literal interpretation of the Bible, ignoring its patent inconsistencies. These are many. One of these, George Bernard Shaw with his incisive genius has pointed out: the evolution of the idea of God as revealed by different Biblical heroes. Other peoples than Christians believe in God, yet they reject the Bible either in whole or in part. Christianity is far from being universal and the Bible far from being recognized as holy by all mankind. The theory that living beings have evolved applies to all mankind and is in this sense independent of any particular religious cult. But it is not incompatible with belief in God.

To set off the sciences as antagonistic to religions is a common fallacy. Indeed, in much of our thinking we err because we subscribe to what I may call the doctrine of false antagonism, false because in it antagonisms are confused with opposites, and contraries not distinguished from either; thus are set up as antipodes what are not such. Moreover, religions of whatever form that have a strong basis, from whence they derive much influence, in natural laws, are often, especially the more naïve, interpretations of natural phenomena. At one point, Christianity, for example, has universal appeal. Here it rests upon a biological foundation. The doctrine of the brotherhood of man is a universal truth not because it is Christian but because it is biological. Biology, if it teaches us anything at all, teaches us that mankind is one—from the most primitive Bushman, lowest human, if you will, to the most perfect, purest Nordic, if nowadays there be such. Whatever the individual conception of human race, as pure and undefiled and as real or as devoid of meaning, biology teaches that, according to its elementary principle—that of production of fertile offspring—all men are equal.

The paramount argument of the thesis stands or falls on this assumption that the world of living things constitutes a series in which we see a nearly linear arrangement of lifeforms in an ascending order. For us, this order finds more reasonable explanation on the assumption of evolution than on that of special creation. For us, subscription to this assumption demands as a logical consequence that we include every quality and every characteristic, all forms of behavior, of

Page 12

man's whole being, physical and spiritual (social) emerged from a low, a simple, and a single fundament which in a progressive development became a highly complex and multitudinous structural and functional composition with manifold relationships. Then ethical behavior also was evolved. But the thesis goes farther: it aims to delineate the very most primitive forerunner of ethical behavior. It thus differs, especially on this point, from previously enunciated theories of the origins of ethics.

Theories of ethics may be classed as those founded on the idea that ethics is a gift directly from God to man alone, and as those not so formed. Of theories on the second class, some imply an evolution; others are vague in their conceptions of the not-divine origin of ethics. One might at first thought regard this classification as identical to one made on a chronological basis: in the first class would fall theories prior to Darwin, in the second those which appeared after the publication of <u>The Origin of Species</u>. Such, however, would not be strictly true. Darwin was by no means the first to espouse the idea of evolution; hence, a classification made with reference to the time when appeared his first work on evolution does not coincide with ours. Further, before the appearance of <u>The Origin of Species</u> ethics was very clearly espoused as the outcome of an evolutionary process. But undoubtedly subsequent to the appearance of <u>The Origin of Species</u>, a great impetus was given to the promulgation of systems of ethics grounded on the idea of evolution. Many authors of these accepted also Darwin's individual contribution, his attempt to explain the cause of evolution.

Among those prior to Darwin's time who derived ethics from God, Hume stands out prominently. He not only postulated such derivation but also insisted upon ethics as the endowment of civilized man alone. This point of view was perhaps justified since man in a state of savagery was not well known in Hume's time. No one today who has even a superficial acquaintance with studies on savage man can any longer sustain Hume's position.

After Darwin, the strongest protagonist for the directly divine origin of man's ethics was, in my judgment, Thomas Huxley—strongest because Huxley was the doughtiest Defender of the Faith (Darwin's) and the most active disseminator of Darwinism, anticipating with eminent success twentieth century methods of propaganda. Indeed no publicity officer of any of our great American universities of today, where a pervading inefficiency cannot damper American genius for advertising, is Huxley's equal in making the man in the streets "science-conscious".

Because he was so thoroughly committed to the concept of evolution, Huxley's essay on ethics, as first read, branded him the rankest apostate. And there are others today in this same category, those who swear allegiance, almost holy, to the theory of evolution and who give unalloyed confession of faith to the Darwinian interpretation of the cause of evolution but do not include man's ethical behavior among his heritages from the past. Strange inconsistency, this, for the only logical position for an evolutionist to maintain is that all of man, physical and spiritual, is the product of evolution. If it be contended that, of man, only his body has evolved, a theory of evolution has neither weight nor value. To exclude man's spiritual being from the purview of evolution, is to exclude man from the process of evolution. Thus Huxley and others who hold similar views apostatize at the very heart of the doctrine.

To be sure, Huxley later revised his famous Cambridge lecture on ethics in the attempt to veil its inconsistency. But the first thought, the second could not undo. Even the most charitable must lodge against him at least a charge of loose thinking.

Now, without doubt, an advocate pleading in Huxley's behalf could stress that Huxley misunderstood Darwin's concept of the struggle for existence, shown by Huxley's quoting of the since then hackneyed Tennysonian line, "Nature red in tooth and claw"— without incorporating which no first-class lecturer ever feels in his heart that he has amply embraced the subject. I would accept the pleading—but not to Huxley's credit or to that of any Darwinian who misconstrues what Darwin meant by the phrase, struggle for existence. Romanes understood it correctly. And so did Kropotkin. Nor would I excuse any Darwinian who would plea that Darwin himself for years allowed to go unchallenged the expansion of this erroneous interpretation.

Page 15

Instead I adjudge Darwin not without fault, despite his clear but twenty years late exposition in <u>The Descent of Man</u>. The insistence with which Huxley fought for the acceptance of Darwinism put him under the obligation to understand that for which he fought. I digress for a moment.

It has ever seemed strange to me that so commonly both writers and thinkers confuse the idea of evolution—what biologists are proud to call the facts of evolution—with the cause advanced to explain the process by which evolution came about. More curiously, the confusion is not limited to non-biologists; biologists themselves, among these the most ardent evolutionists, too frequently fail to make this distinction between "facts" and "cause" to explain them. They easily forget that evolution and Darwinism are not synonymous. Most biologists accept the postulate of the "fact" of evolution; few agree concerning one or another given answer to the question, how evolution came about. With the late William Bateson many present-day biologists agree: the cause for the evolutionary process has yet to be determined. Hence, the Lamarckian concept of the inheritance of acquired characters is no more barren than the Darwinian theory of natural selection. And the latter, no less than the former, invokes the influence of the environment.

Perhaps the immediately foregoing is not so much a digression after all.

Page 16

For any thesis of man's ethical behavior whose hypothesis is evolution, it is obligatory to distinguish between the process of evolution and that which brought about the process. Failure to make the distinction when a subject like ethics is broached is more serious than when the anatomy, physiology, taxonomy or other laboratory (and field) discipline of plants and of animals is apprehended by observation and experiment. The more tangible and phenomenal the subject, the less serious the failure; the less tangible and the more non-phenomenal, the more serious the failure to differentiate between the "fact" and the "cause", the failure may not be attributed to the non-validity of evolution as "fact." Failure to sustain an ethics based on the Darwinian concept, as Huxley understood it, "Nature red in tooth and claw", in no wise hypostasizes evolution as "fact" I would go farther.

Whilst evolution does not lose if a theory of ethics built upon a particular causal theory of evolution fails, this failure would, to me, indicate the inadequacy of the causal theory. For me, the efficacy of any theory of the cause of organic evolution is measured by the degree to which it is capable of sustaining the superstructure of a theory of the origin and evolution of man's ethical behavior.

To return to the main line of the argument. Hume and Huxley by no means exhaust the list of those for whom the source (the direct source) of ethics is the Creator. Many philosophers belong here and, of course, most of the Christian Clergy. Hume and Huxley have been singled out because of the relevance of the fallacy and the inconsistency respectively of their essays on the present thesis.

Expositors of theories of ethics of not-divine origin, although not so numerous, include names as illustrious as those in the class first named: Shaftsbury, Proudhon, John Stuart Mill, Herbert Spencer, Darwin, Guyau and Kropotkin, among others. Whilst some of these were not either clear or consistent as to origin or their systems, Guyau and Kropotkin clearly were. These, one a gifted short-lived philosopher, the other a rare combination of biologist and thinker, I follow in no small measure in the development of my conception of the origin of man's ethical behavior as the product of evolution. Perhaps had Kropotkin lived to complete his <u>Ethics: Origin and Development</u> [1924], I need not have written this book. On the other hand, since he did not have at his disposal the evidence on which my conception is based, it may be doubted that he could have expressed the central idea of my book. I do not aim to undertake the task of re-stating Kropotkin's thesis; it does not need any restatement. Rather, I propose to go farther than he went—to trace to its beginning in the most primitive form of living substance that from which is derived what I define as man's ethical behavior.

Page 18

Chapter 2 - **EVOLUTION: OF FORM AND OF FUNCTION –** pp. 18-35

Although Darwin's theory of natural selection as the explanation of the evolution of animals and of plants is far from being accepted by any substantial number of biologists, his assembly of the evidence in support of the proposition that evolution is the mode by which the forms of life came into being, not only is widely accepted by biologists, but also elicits almost unanimous assent among them. Other thinkers as well, outside the field of biology, estimate that this mass of evidence preponderates that for the theory of special creation. Before Darwin, as early as the time even before Aristotle, men had guessed more or less vaguely concerning evolution, with or without venturing to a tentative cause. But the most anti-Darwinian cannot fail to admit that, natural selection aside, Darwin was the foremost in assembling with expert craftsmanship the evidence of fact <u>for</u> an evolution. As would the creator of a symphony bring together folk-songs, the instinctive expressions of a people's common heart, and weave them into a magic pattern of his own to move again this heart, so Darwin gathered the vague ideas long current, the tentative speculations, the ungarnered acts and made of these a harmony.

Whilst I in no wise ascribe either to the theory of natural selection or to the most prevailing interpretation of what Darwin meant by the struggle for existence—that which makes

Page 19

"Nature red in tooth and claw" [Tennyson], where wages perpetual internecine war and death except for the most brutal is the beneficent and quick surcease for an all-cruel life—I do not hesitate here to review this evidence. This I rehearse briefly; it is too well known for extended exposition. Since I deal with the origin of a human manifestation, I limit myself to evolution in the animal kingdom.

The sources of evidence whence are derived facts for support of the evolution-theory are in the following sub-divisions of zoological study: of fossil forms, of geographical distribution, of the classification of animals, of comparison of structures, and of the development of individual animals.

Fossil forms. The study of fossils (Paleontology) permits the statement that in rocks is a record which indicates that forms of life appeared successively in successive periods of the earth's history. For various reasons, one cannot consider this record as complete. For example, in the depositions, mainly hard parts alone of animals were preserved. Thus were preserved tests of organisms, including some protozoa, sea-urchins, star-fishes, etc.; exoskeleton of crustacean, insects and their allies; shells of clams, snails and the like; and armors of reptiles and other vertebrates. Soft parts of these creatures and wholly soft-bodied organisms were less well preserved. Nevertheless, there are fossil remains of these. The late Professor William Patten, a well-known paleontologist of rare skill, had in his collection specimens of now extinct species of fish in which one could discern the course of the great nerves connected with the brain.

Study of these fossil remains contributes very strongly to the theory that in the history of the earth animals appeared successively. Each era had a characteristic fauna; in these a group of animals appeared for the first time, whereas another became extinct never to appear again. Up to and including the Devonian, a period in the Second Geological Time, the Mesozoic, fishes alone represented the vertebrates. Successively amphibians, reptiles and mammals appeared.

Geographical distribution. When one compares various regions of the earth's surface with each other with respect to the distribution of the fauna, one notes that the species of animal forms are different. It is well known that the fauna of South America, Africa and Australia differ. Wide seas separate them and they have been separated for ages during which the fauna on each has developed in a peculiar way, independently in each region. Australia, longest isolated, shows most primitive mammals except those that could by flight reach it or the smaller ones brought in by driftwood. South America and Africa have been so long separated that in each the fauna was able to develop along different lines. Even in regions more nearly adjacent and subjected to the same climate, barriers sufficient to restrain migration or dissemination isolated some of the fauna which developed along different ways.

Page 21

Islands also are interesting. In the British Isles which were late in separating from the mainland and not far away from it, the fauna comprises few forms peculiar to these islands. Japan, earlier separated from the mainland of which it was once part possesses more. Madagascar having still earlier become an island has a very peculiar fauna. These and other facts of geo-graphical distribution argue for evolution.

Classification (Taxonomy). Animals may be arranged in groups which are mutually exclusive. Thus, all radially symmetrical animals with spiny skin which live only in the sea and develop from bilaterally symmetrical embryos, are echinoderms—sea-urchins, star-fishes, sea-cucumbers, and the like. No other animals can be placed in this group. There are other radially symmetrical animals in the sea and in fresh-water, the jelly-fish, comb-jelly, coral, etc.; but they are, in the light of our knowledge, classified as formerly with the echinoderms. Similarly, all animals that have jointed appendages, an outside covering made up of chiton—a material which contains an interesting chemical, glucosamine—which is periodically shed, are arthropods: crabs, spiders, insects, etc. Vertebrates have jointed appendages and some, the turtles, for instance, have an exoskeleton; but the appendages are of different organization and the turtle's exoskeleton is chemically, histologically and embryologically unlike that of the lobster. So again, all animals, with or without a backbone, that possess a notochord, the rod around which the back-bone forms when present, are chordates. In addition, for each group are other diagnostics that define the group as different from others. All the foregoing is true for twelve great groups or Phyla each of which is sub-divided into classes, orders, families, genera and species in turn—omitting the sub-divisions which in the hands of the extremist in taxonomy may go on ad lib. It should be pointed out that there exist some animals of uncertain affinities; these are placed in one or the other of two phyla of near position.

Now, when we examine these phyla more closely, we ascertain features that allow us to arrange the phyla in an order on which in the main zoologists are agreed. This arrangement places the single cell animals first. Above these are the first of the many cell forms, the sponges, and then the jelly-fishes and their allies; their cells are arranged in two layers, endoderm and ectoderm. The flatworms are next, followed by the round-worms; in addition to endoderm and ectoderm, they possess a third layer of cells, the mesoderm, but lack a body-cavity (coelom) lined with mesodermal cells. Above these acoelomates, are the coelomates—animals in which mesoderm cells enclose the gut and line the inner side of the body wall. Within the coelomates one recognizes grades of complexity such that one arranges them in an ascending order culminating in the twelfth phylum, the chordates, with its large sub-division of animals with backbones. Among these, the vertebrates, is again an order of increasing complexity, culminating in its final term, man.

Page 23

Taxonomy depends upon knowledge of form, which often means dissection. Hence, comparative anatomy and taxonomy are auxiliary. Comparative anatomy yields further and strong evidence for evolution.

Comparison of structures. If we compare the skeleton of upper and lower limbs of man and frog, we are struck more by resemblances than by differences. In each the upper, pectoral, limb begins with the pectoral girdle from which depends the bone of the upper arm joined to which are the two bones of the lower arm, and ends with the hand and the fingers. Likewise the skeleton of the pelvic limb: pelvic girdle, bone of the upper, two bones of the lower leg, and the bones of the foot and the toes. The pectoral and pelvic fins of a fish have the same structural ground-plan. Every vertebrate has a heart, arteries, capillaries and veins through which courses blood carrying white and red cells except mammals in which the red cells are replaced by pieces of cells. Or examine the digestive, uro-genital or nervous system: these show in vertebrates the same fundamental plan of organization. The respiratory system, despite the presence of gills in some and of lungs in other vertebrates, exhibits sharp resemblances. For all the organ-systems one concludes, on the basis of the mass of evidence, that from the vertebrate of lowest taxonomic position to man, there is a progressive increase in complexity of structure. Thus, it is true not only for the phyla themselves but also for classes within them that one recognizes an order which indicates an evolutionary process.

Page 24

The development of individual animals. That sub-division of the science of zoology, embryology, which concerns itself with the development of individual animals from eggs applies to member of eleven of the twelve Phyla. Members of the first Phylum, the Protozoa, do not pass through a process of embryogenesis since they have neither embryos nor eggs. Inasmuch as this statement may be denied by some specialists of the Protozoa, it may be amplified.

The sole method of reproduction among the Protozoa is that of fission, division of the cell into two or more, equal or unequal, fragments each of which as a new individual repeats the process anew. For many Protozoa it seems necessary that there be at some time a fusion of two individuals; otherwise, the life-history ends. This fusion may be between two like or two unlike cells; in the latter, the partners may show so great disparity in size that they resemble the egg and the spermatozoon of a multicellular animal, a likeness all the more striking since the smaller cell may be highly motile.

For Metazoa also fission is the mode by which a new individual comes into being with, however, the important difference that it is always the egg that undergoes the fission. Even for those Metazoa as larval forms or as adults that have the power of breaking themselves up into fractions, each fraction constituting a new individual, it is none the less true that reproduction is by means of fission of the egg. If, as some students of the Protozoa claim, every protozoan must at some time or another intercalate fission with a fusion of cell with cell in order to maintain the power of reproduction, I would not agree with the proposition that this fusion, even between cells of wide disparity with respect to size, to motility and hence to organization, is a type of fertilization similar to that involving the union of egg and spermatozoon as in Metazoa.

First, what some protozoologists denominate fertilization involves two entire organisms. This fact alone so strongly militates against the definition of fertilization that it becomes untenable. Eggs and spermatozoa among Metazoa are always parts of the parent or parents. And if those who insist that union of two protozoan individuals of marked inequality in size is the same phenomenon as that of metazoan fertilization, because of this size-difference, they argue against themselves when they, as they do, define as fertilization the union of two similar protozoan cells. Among Metazoa, marked disparity of the partners with respect to size is the rule.

Secondly, granting again what has not yet been proved, that all Protozoa must at some time in their life-history undergo pairing, we note another difference to the Metazoa. This is that for those Protozoa in which fusion of unlike individuals has been demonstrated to occur, failure of fusion does not terminate life. Thus the protozoan that causes malaria fever can go on in the mosquito or in man despite the failure of union of the minute spermatozoon-like individual with the larger egg-like one.

Page 26

One might here raise an objection since some multicellular organisms, flatworms, for example, may go on indefinitely reproducing themselves by fragmentation. This is true. But we must recall that such forms began as eggs. The large egg-like form of the malaria parasite arises in response to an extreme change in the environment and is so to say a specialized adaptation. The animal egg is the very ground-work of metazoan reproduction.

Some Metazoa, the honey-bee, for example, produce eggs that develop without fertilization. In this process of the initiation of development, parthenogenesis, without the intermediation of the spematozoon, the egg is no less an egg. The phenomenon of parthenogenesis brings us to consider the third point.

In the case of a large protozoan individual that resembles an egg, the fusion of this individual with the spermatozoon-like cell depends on the ensuing phase of the life-history. Yet, for no such egg-like organism have claims made for its parthenogenetic development ever been accepted by protozoologists generally. Among some Protozoa occurs a process of internal rearrangement, called _endomixis,_ in contrast to _amphimixis_ which involves rearrangement due to introduction into the egg of material by the spermatozoon in fertilization. But endomixis is not comparable to parthenogenesis in an egg—for several reasons. One may be given. Endomixis occurs in an organism in default, presumably, of its pairing with another. Since in such cases the individuals that would pair are alike, endomixis probably occurs in either, or in each. But, so far as I know, there is no observation on record among Metazoa of the

Page 27

parthenogenetic development of a spermatozoon.

Only by misuse of terms can one speak of embryos in connection with Protozoa. For embryogenesis, the egg is sine qua non. For the three reason given—and others could be—we conclude that Protozoa do not produce eggs. Protozoa are without sex, although because of sexuality being present, they foreshadow sex, whose fundamental diagnostic is the production of eggs and spermatozoa.

It is a misfortune that some zoologists overlook these facts: by extending the term, fertilization, union of egg and spermatozoon, to include phenomena met with in Protozoa that simulate those in Metazoa, they have masked a most intriguing question, one the answer to which might furnish contributory evidence in support of the evolution-theory.

Striking is the well-known fact that every Metazoan, from sponges to man, begins its life-history as an egg, fertilized or not. To be sure among these are forms which have the capacity for asexual reproduction – by fragmentation, bud-formation, etc.; but this mode never stands alone but rather alternates regularly or by accident with reproduction from an egg. Thus, egg-production is the property of Metazoa only, and it is the sole property common to all Metazoa. The production of spermatozoa also is found exclusively among Metazoa; but spermatozoa themselves are in final terms derived from eggs.

Page 28

Every animal egg pursues the same course in preparation for its initial stage of development into an embryo. This consists of two divisions whereby three-fourths of its nuclear substance is lost and most of its cytoplasmic bulk is retained. Although these divisions (of maturation) in some eggs ensue wholly or in part before fertilization or the exciting means for eliciting parthenogenesis and in others only after the one or the other stimulus, they obtain in all eggs. There is no correlation of the time of maturation with respect to fertilization and the position of the animal in the scale of animals. Spermatozoa without exception pass through two divisions before being prepared to incite the egg to develop; only sperm-cells that have completed maturation and undergone certain cytoplasmic changes ever fertilize eggs. Thus the eggs, the products common to all Metazoa, show with respect to nuclear organization in the initial stage of development the same picture.

The developmental process of all eggs begins in the same way—as a response by the egg surface to the spermatozoon in fertilization or to the means, natural or experimental, that elicits parthenogenesis. Again, therefore, the common product of Metazoa reveals the same behavior. Indeed, this is in some cases striking, for in such, the presence of the spermatozoon necessary for this surface response by the egg, having elicited it, takes no further part in fertilization as in the majority of cases. But the ground common to eggs' development does not end here.

Page 29

Once stimulated, every species of egg goes through a process of self-sundering, a breaking up into many cells. However much this cleavage may vary with respect either to the pattern of adhering cells produced or to the time after fertilization when demonstrable cleavage-planes become visible, cleavage obtains in all eggs. The surface-changes of an egg at the time of fertilization are visible manifestations called forth by the spermatozoon, and they announce the elevation of the egg's life-processes to a higher level. In a similar though less dramatic fashion, every stage in the cleavage-process exhibits a characteristic surface-behavior. By this also the cleavage-cells are en rapport through the spinning of delicate threads which maintain the union of the cleaved mass.

Finally, every single animal egg, thus cleaved, forms of the cleavage cells two layers, the ectoderm and the endoderm. Ectoderm and endoderm differ visibly in their organization and inherently, as their subsequent history reveals. A third layer of embryonic cells, the mesoderm, found in all Metazoa beginning with the flatworms, cannot be considered as common to all egg-producing animals.

These well-known facts have not been hitherto brought together in this wise. And they should be emphasized, especially by students of evolution. Instead, among the strongest protagonists of evolution are eminent embryologists who insist that every egg is a law unto itself. This is tantamount to saying that every egg has been specially created!

Page 30

On the characters common to all eggs are superposed specific qualities and manifestations and on the early phases of embryogenesis common to all egg-producing animals are superposed those features which indicate that animals have further evolved. But it is to the common bases enumerated above that I would call attention. In the light of the foregoing argument: how may we attempt to explain the sharp dichotomy in the animal kingdom of Protozoa and Metazoa—i.e., with respect to the capacity of an organism to split off part of itself for reproducing its kind?

If we knew beyond question all that there is to be known concerning the evolution of living forms, we should scarcely need postulates, hypotheses—theories even. We are far from having at hand sufficient tangible evidence from which to derive such knowledge. Evolution-theory, the bed-rock of modern biology, is thus a concept. Although most plausible, it is as yet incapable of laboratory experiment. There is therefore still necessity for speculation concerning the process of evolution. In default of fact concerning the origin of the first egg-producing organism, we can only offer suggestion.

Let me begin with the generally accepted proposition that the principle of the division of labor plays a role in animal activity—a point which I discuss beyond in another connection.

33

In a single unicellular organism, Amoeba, for instance, there is no division of labor comparable to that in an organism made up of many cells. In Amoeba, although there are loci for the inclusion of food and for eliminating excreta within the cytoplasm, and although the cell surface takes the lead in locomotion, there can be no division of labor in the sense of the distribution of work to the members of a community. Even in a highly developed colonial protozoon, division of labor does not attain the degree reached by sponges; the organization of such a colony is not similar to the individual sponge. At the very least one must admit a greater degree of division of labor among multi-cellular than among unicellular organisms.

Now, in the course of evolution, division of labor we may assume played a part. Increasing complexity demanded more and more exquisite means for dividing up work. Or, more refined division of labor called forth greater complexity of organization. A protozoan cell must be a tremendously complicated structure of protoplasm—if not the most, the second most in the animal kingdom. But the activities confined within its single boundary presumably are checked by this singleness. And the very complexity in this so little space is a further check. Protozoa did not expand therefore beyond a certain limit. Their evolution, that is, became circumscribed by the singleness of their cellular make-up.

Page 32

But somewhere along the line of the evolution of the Protozoa—or from the Prototype of both Protozoa and Metazoa—emerged what was to be the first Metazoan. I suggest that the greatest step made in the whole evolution of the Metazoa was that by which the egg was produced; so great indeed that, once taken, all farther advance was made on this basis common to all Metazoa, egg-production. Among the Protozoa is no division of labor with respect to maintenance of the individual and the reproduction of the species; the single organism is both field for maintenance of self and substratum for the continuity of its kind. With the Metazoa enters a sharp division between the cells of the individual life and those for the perpetuation of the species. Moreover, the period of the individual's growth, so short and bound so closely to reproduction among the Protozoa, lengthens and becomes farther removed from the egg, the substratum of reproduction. This holds no less for those Metazoa which in their larval stage produce eggs.

This difference may be expressed otherwise. In a Protozoon, nutrition and the whole of the catenary processes of individual life leading to growth are visibly subservient to the continuity of the species. In the metazoan this subordination is so masked that the individual has two lives, one selfish and one for the species. But note also a common feature: Protozoon and egg are single cells. And if the former is the most complex cell in the animal kingdom, the latter as single cell approaches and in prospective outcome surpasses it. Thus although this step taken in evolution when the first egg-producing form arose was tremendous and new, its starting-point, cellular organization, was old. This start set the way along which the evolutionary process moved.

A physiological process apart from underlying structure is unthinkable; study of structure and of function are therefore two aspects of zoology which, though for convenience are separately studied, allow for no sharp lines of demarcation. Hence, there has been an evolution of function as well as an evolution of structure. Cellular organization, that of the protoplasmic mass into nucleus and cytoplasm, once having been attained, became the structural characteristic common to all later appearing animal organisms. The cell, unit of structure in cellular organisms, is also the unit of function. The evolution of function is thus to be sought in the display of activities by the lowest form of cell and the compounding of these to constitute the manifold exhibitions of behavior that characterize the more and more structurally complex multicellular organisms, culminating finally in human behavior.

The evidence for the evolution of function in whole organisms, in their systems, in the organs of systems and in the tissues that make each organ, is sufficiently well known. This evidence frequently reveals relationships between forms which outwardly do not closely resemble each other. Thus the semi-solid urine of a reptile and bird indicating a similarity in function with respect to retention of water in renal activity, points strongly to close affinity between the highest of the cold-blooded animals and the first warm-blooded. The vital functions of birds and of mammals, possible only within a rather narrow range of temperature, indicate the kinship of birds and mammals. The similarity of chemical constituents found in the urine of apes and of man is an expression by function of near relationship. Respiration is in all the vertebrates mediated in part by the red blood corpuscles; for they transport oxygen to the tissues, a transport due to the presence of hemoglobin. Hemoglobin can be crystallized. The resemblance of crystalline structure is closely correlated with the taxonomic position of the animals whence the hemoglobin crystals are derived. In these and in other ways one can trace an evolution of function in every system, organ and tissue of the human body.

In the final analysis, however, the work of a multicellular organism is the sum-total of the work of the single cells that build it up. All cells, protozoan and metazoan, and plants as well, are possessed of certain common displays of function. These are respiration, nutrition, contraction and conduction. Respiration in all living things is the same—an exchange of oxygen and carbon dioxide. In nutrition most animals and most plants differ. Contraction and conduction are more emphasized in animals than in plants. Conduction is the business of the

Page 35

nerve-cell. Hence, the nervous system of all higher animals that possess such is primarily an elaboration for conduction. Man's nervous system, structurally and functionally, sets him apart from all other animals.

Page 36

Chapter 3 - EVOLUTION: THE HUMAN NERVOUS SYSTEM – pp. 36-51

Man's nervous system comprises a central structure, the sense organs, and a network of nerves. This whole intricate system, by virtue of its ramifications which extend to all parts of the body, differs from all other systems of the body. The blood and lymph systems also extend to all parts of the body, it is true, but with the significant difference that their extensions are means for transport of media for chemical integration whereas the nerves themselves are actively engaged in assuring by physical contact the interdependence of the various body-regions on the one side, and by connections with the sense organs, relation to the environment, on the other. This complex structure of the nervous system, as every other structure of the human body has evolved. Evolution of nervous function can also be traced.

This central nervous system consists of spinal cord and brain. In the cord when cut across one notes near the center a minute hole, the central canal in a section, surrounding which the substance is grayish due to the preponderance of cells in this area. Enclosing the gray substance is the white, made up of cell-prolongations or nerve fibers. The nerve-fibers are arranged in bundles or columns that extend lengthwise along the cord. Some are near the top, some near the bottom, and others at the sides of the gray substance; these are the dorsal, ventral and lateral columns, respectively. The dorsal columns are composed of fibers that enter the cord from cells which are located in knots, dorsal root ganglia, one to each segment of the spinal cord; these are the afferent nerves. Efferent nerves have their cells of origin in the ventral horns of the gray matter and emerge from the ventral side of the cord. Afferent and efferent roots are bound together beyond the size of the spinal ganglion. This arrangement, of gray and white substances, is common to all vertebrates; the variations of these substances from lowest to high are such that they indicate a progressive evolution from simpler to more complex. The spinal nerves enter and leave the cord in the same way in all vertebrates except in the lowest, the fishes without jaws, lamprey, for example, where the dorsal and ventral roots enter and leave the cord alternatively instead of in pairs.

The human brain, as that of all vertebrates, is subdivided into five regions: primary and secondary fore-brain, mid-brain and primary and secondary hind-brain. Evolution of the brain is shown by the gradual reduction of some regions and the gradual increase of others. From fish to man the olfactory lobes diminish in size. In man, the pineal body, which in the lowest fishes ends in the pineal eye located on top of the head, is reduced to a process overgrown by the cerebral hemispheres. In the lowest vertebrates, the cerebellum is a mere flap of nervous tissue; in birds and in mammals it reaches its greatest development. The outer layer of cells in the cerebral hemispheres deserves special mention. Beginning with the lowest forms, this layer of cells, the pallium or archipallium, is very thin. Progressively in the ascending classes of vertebrates, the pallium thickens; in mammals it is folded; the degree of this involution reaches the highest state of development in man.

Now in this pallium, or cerebral cortex, can be demonstrated certain areas of cells called sense, motor and association areas. The first named are centers for the various senses; the second contains centers for motor activity. The association areas constitute areas the cells of which establish connections with other areas.

The nervous system of vertebrates arises in the same way, from a flattened plate of superficial cells that extends lengthwise to the right and left of a median line along the dorsal surface of the embryo. By apposition of its lateral edges, this, the medullary plate, becomes the medullary tube. Anteriorly, the tube becomes the brain; posteriorly, the spinal cord. The tube sinks and is overgrown by ectoderm cells. The medullary tube in no vertebrate ever becomes solid but retains a lumen, the central canal of the cord and the ventricles of the brain, the cells adjacent to which possess cilia. Thus what were once the most external faces of these cells before involution of the medullary plate to form a tube now line the canal and the ventricles.

Page 39

The development of the nervous system may be followed in embryos of reptiles, birds and mammals, including man. To trace it with facility in living embryos, eggs of fishes and of frogs, toads and salamanders are to be preferred.

The eggs of common frogs, when laid, show one hemisphere brown or black and the other white or cream-white, depending upon the species. In the course of twenty-four hours after fertilization, the eggs are almost entirely dark, due to the overgrowth of the lighter hemisphere by cells that contain the darker pigment. This overgrowth continues until all but a plug of the more lightly colored cells protrude. The plug (the blastopore) in turn is obliterated by the overgrowth of the more heavily pigmented cells. In front of the blastopore, the egg becomes markedly flattened, an indication of the medullary plate. The lateral margins of the plate arch up to form the neural crest; the plate sinks along its length to form a gutter, producing three out-pouchings anteriorly—the fore- mid- and hind-brain vesicles. Gradually, the margins of neural crests approach each other, bringing about the formation of a tube from this gutter with arched sides. The tube sinks below the surface and the cells of the body wall cover it. Much the same process takes place in the egg of a fish. To follow under the microscope the development of the nervous system in a transparent fish-egg is an engaging spectacle.

Virtually, the structural ground-plan of the vertebrate nervous system is identical for all vertebrates, for although differences obtain, these are to be correlated with structural characteristics peculiar, first broadly to the vertebrate classes, next more narrowly to orders in each class, and finally to genera and to species. Thus one should be ever wary to interpret on the basis of the investigation of processes found in one order processes found in another. For example, the development of the neural crest in salamanders differs from that in tailless amphibians. But so strict is the adherence of the development of the vertebrate nervous system to the ground-plan as outlined above, that we may say that this system has been cast in the same mold. Upon this primitive foundation the vertebrate nervous system has developed from lowest to highest with increased structural accentuation in some regions and decreased in others. But always the brain and spinal cord arise from ectoderm cells, which become dislocated from others. Although they come to be located within the body, they are derived from cells of the surface of the embryo.

The entire nervous system is a network of individual cells of ectodermal origin and comprises cells of nervous activity, the neurons, and supporting cells, the neuroglia. Structurally, the neuron is characterized by two (or more) processes, axone and dendrite. Primarily by means of these processes, the neuron, and hence the whole nervous system, carries out its chief function, conduction. Broadly considered, this is integrative in two ways: of the parts of the organism and of the organism with the outside world. By the former, a nicer adjustment of the interdependent

Page 41

organ-systems is maintained, and thus a more exquisite performance of divided labor; by the latter, the organism is brought into more acute awareness of the external world. From lowest vertebrate to man, these two aspects of nervous function may be traced, paralleling the evolution of nervous structure. Whether a nervous action integrates parts of the organism or brings the organism en rapport with its environment, the functional unit is the neuron in association with one, two or more neurons.

A nervous reaction involves a) the reception of a stimulus, b) the transmission of the stimulation to the brain or cord, and transmission from center to the region where, c) the effect is evoked. The simplest possible action of this sort demands at least two neurons: the end of one, beginning at the receptor, picks up and transmits the impulse over the conductor (nerve-fiber) to a cell of origin in the spinal ganglion or in the brain which cell by means of its other process conveys it to another neuron whose axon conducts it to the effector. This, the simplest reflex, is possible, although it probably does not exist among vertebrates; nevertheless it serves to illustrate the anatomical basis of a reflex arc.

The reflex arc is often more complicated, calling into play within cord or brain two or more neurons intercalated between afferent and efferent neurons. A further source of complexity resides in the rich connections rendered possible by the many branches of the cell-process. Thus an axon may present several collaterals, extensions almost at right angles to the axon; by way of these impulses coming from various body-regions may be discharged to the cell of an efferent neuron. In such cases, however, the path from center to effector is final and common, always the same, no matter from which of several receptors the initial impulse originates. Thus the reflex arc may be simple or much compounded. It may also be coordinated, in a chain. It may elicit another response antagonistic to it. Its center may be wholly spinal or cranial; it may have by connections centers in both.

Very intricate movements can be demonstrated to be purely reflex. A frog with its brain severed from cord will continue to perform acts that one would at first hand consider possible only through the mediation of the brain. The "spinal" mammal, i.e., one whose brain and cord are no longer in connection, will behave similarly. These laboratory experiments evince the supposition that centers for reflex action are located in the spinal cord. One should however recall that the "spinal" frog that swims and the "spinal" cat that scratches itself, thus acting in a purposeful way, perform these movements only when stimulated. It would be a mistake to conclude that in the intact animals swimming and scratching ensue without the intermediary action of the higher nerve centers. A human syphilic whose brain is diseased may not be able voluntarily to move his arm, which

Page 43

does move when directly stimulated. Likewise, in the condition known as Argyll-Robertson pupil, the iris of the human eye accommodates from darkness to light—i.e., the pupil becomes smaller—whilst the parallel change for far and near vision does not obtain. These examples teach us that on reflex actions are superposed the voluntary.

It is well to recall that the brain also is the center for reflex actions. Some of these, the vital reflexes indispensable for life, are located on the floor of the fourth ventricle (in the medulla oblongata, the hindermost region of the brain). We do not will the heart to beat or the respiratory movements to take place. Nor can we _will_ to alter a reflex, although what were originally reflexes may in some cases come under the will. A reflex action may be conscious, contrary to a rather widespread notion that classifies reflex actions as unconscious. Finally, nervous actions are either voluntary or not. If they are involuntary they are reflex.

There is not one kind of complex of nerve-fibers for a voluntary action and another for an involuntary; the _modus operandi_, the structural basis, are the same for each with one important difference. This is that the performance of a voluntary act demands the intermediation of cells of the cerebral cortex. Thus, a willed act differs only in having a localized center; otherwise, a reflex arc may serve also to illustrate transmission of the impulse in a voluntary act. Thus, it is in the cerebral cortex, in its structure and its functions that we seek the differentia which separate man not only from the lowest vertebrate but also from primates nearest him. These have come about through a long process of evolution involving millions of years; this we conclude from study of the comparative physiology of the vertebrate brain. Study of physiological processes occurring during embryo genesis and of those in early infancy support the conclusions derived from comparative physiology.

Briefly, the evidence from comparative physiology indicates that, from lowest to highest vertebrates, the functions of the cerebral cortex (as well as other parts of the nervous system) have steadily gained, culminating in an intellectual activity which separates man from all other living forms. More and more, the highest nervous functions have come to have their seat, so to say, in the cerebral cortex, in areas having been fairly well marked out by experiment and by studies of diseased brains. The data for the embryology of man's higher nervous functions, although not so numerous, nevertheless indicate that they too develop progressively. Finally, for the human infant are lines of evidence that permit us to conclude that functions associated with the cerebral cortex are gradually established.

Since some reflex actions are conscious, we cannot say that the consciousness of actions is the exclusive property of voluntary acts. The consciousness or the being aware of a reflex does not, however, imply that in the completion of the action

Page 45

the cortical (cerebral) cells take a direct part. Nevertheless here is indicated a liaison between reflex and voluntary action. Additional kinship, on the side of structure, as has been shown above, appears in the utilization of the same morphological track, neuron to neurons complex, by both reflex and voluntary actions—with the exception that the voluntary action demands the intercalation of cortical cells. Thus, the two types of action are closely similar. And when we add to the simplest voluntary action a compounding by the addition of more and more cerebral cells—those especially of the association areas—the initiation of voluntary action by the brain itself and the discrimination of impressions sent in from the sense organs, we but compound the schemata of purely reflex acts. Thus reason and thinking have a morphological texture comparable to reflex.

Reason and thinking we relate therefore to the highest brain centers; their morphological substrate is a nexus of cerebral cells. By reason, by power of the intellect, we say, the individual puzzles out a new situation. This capacity is linked with memory of personal experience or of the experiences of others remembered by the individual. In this sense, a situation is not always new. Others more nearly deserving the application, new, as discoveries and inventions, are new only in their end-points reached by thinking and reason, by the cognition of experiences. Reason and thinking are exclusively individual differentia even when they relate to the experience and ratiocinations common to others. By his reason and thinking, every man tends to isolate himself; for, I repeat, the individual's cerebral processes on the most common phenomena, and generally accepted explanations of them are never the same as those of another. Reason constitutes the last attribute handed on to man in evolution and expresses itself in the manifestation of the "I" or the "Me" in every individual. This self-predominance as revelation of reason, and reason itself, rest upon a broader basis. For, having accepted the evidence in favor of evolution, we accept also that reason has evolved.

Instinct, however defined, is the precursor to reason in much the same way that reflex action is precursor to voluntary. Let us say that an instinct is a bundle of reflexes involving no volition on the part of its portrayer. I would in addition insist that reflexes per se are not thereby instinct. A reflex act demands an initial stimulus; this may come from the outside world. On the other hand, many reflexes begin and end wholly within the organism. If these did not exist, the organism could not maintain life. In vertebrates, the involuntary nervous system, sending and receiving messages from the viscera and central nervous system, is given over to these internal reflexes. I cannot conceive an instinct as an internal reflex or reflexes. If the action is such, how could we observe it? The diagnostic character of an instinctive action is its recognition by the observer: it is out in the open, so to say.

Page 47

This visible display of behavior permits the assumption that, in instinctive action, organism and environment are integrated. Whatever the source for its initiation, within the organism or without, the response is by way of an exhibition of organism and environment as an integral. A further characteristic is consequent: instinctive behavior involves the entire organism; the stimulus (or stimuli) evokes a reaction that pervades, irradiates, its entire being; the organism in its whole responds by active translocation. Passivity of the entire organism in instinctive action never obtains. To be sure a strictly localized reflex act may induce movement; but in the truly instinctive behavior the observer cannot state that its induction springs from one and only one stimulated area. Stimulation is diffuse.

The young sea-turtle leaves its nest on the sea-shore after hatching and without question of aye or no moves straight into the sea. Up to now, no really satisfactory explanation of this instinct has been proffered. We may offer the explanation, or the confession of ignorance, that this behavior is predicated upon the intrinsic organization of the animal. This, and instincts in general, we interpret as the reporting of an inherited interdependence of organism and environment, an interdependence never relaxed and which here reaches a stage of exaltation for a generalized behavior. Instinct is such both with respect to stimulus (or stimuli) and to response. From it, more definitely spatial and particularized behaviors arose during the course of evolution.

Page 48

This interpretation dismisses some notions concerning automatism, acquired instinct, etc. Automatic action, as the heart-beat is such because of the intrinsic qualities of the cells involved; these begin to contract before nerves reach them. The action of the nerves on the heart is to inhibit and to accelerate the beat; it is a reflex action regulating an already existing automatism. It is otherwise with the instinctive behavior of an entire organism.

Voluntary actions are often improperly designated as automatic. The playing of a pipe-organ by an experienced performer, using legs, feet and fingers while reading or remembering the notes can in no wise be considered automatic; thought is involved. Practice has brought about such a reduction in time between thinking and doing that the process appears to be automatic. A skilled needlewoman will knit without regarding the needles, in the meantime carrying on a conversation—to me almost as complicated a performance as playing the pipe-organ. But this and even simpler performances are not automatic. And they are certainly not instinctive.

Instincts are often catalogued as innate and acquired. In the light of the foregoing exposition, we reject the proposition that there exist acquired instincts. Instincts are not individual qualities; they are as much species-diagnostics as are structural characters.

Page 49

It is sometimes said that instinct is tinctured with reason. This is true but not invariable. In such entangled behavior it might be possible to separate the components. The existence of such behavior does not imply that reason is an invariable concomitant of instinct. The evidence permits the statement that the brain is the seat of man's highest mental faculties. The "mentality" of vertebrates then exists in proportion to the presence of cerebral areas comparable to those where reside the emotions, reason, intelligence and thinking—the mentality, in short—in man. In the absence of these areas, the vertebrate's actions are reflex of which some, taken together, are the substrate of instinct. Finally, animals exist that do not possess cerebral structure; in them instinct is highly developed. For the purpose of clarity, we adhere to the definition of instinct as an inherited, involuntary and unconscious action involving the whole organism which responds visibly and <u>in toto</u>. Thus instinct is an "organism-as-a-whole" behavior.

Now, reason and instinct resemble each other in several ways. Both use the reflex arc as a track of operation. Both integrate organism and environment. Each involves the whole organism. But mark the differences!

In the first place, reason is conscious activity. Instinct, we say, is blind. Although it attains an end, this is not previewed. Only to the observer is an instinct-action teleological.

Page 50

The insect that lays its eggs in the body of a species of caterpillar has no prescience of the outcome—less have the larvae hatched from the eggs, which go directly to a fixed locality for further development. These are amazing feats but we do not assume for them conscious rationalization.

In the second place, the integration of man and environment by reason is no longer, as instinct, on the plane of the material. Reason opens up, not new vistas, but those hitherto hidden; but these, received now by highly developed sense-receptors, convey messages that are sorted, discriminated, stored and then ordered again. In this wise not only are new environmental conditions met, but also another environment is opened. Reason compounds itself, grows and extends to the outermost limits of time and space. It is not bound, as instinct is, to the monotonous track traversed during thousands of years by forbears. Reason is self-perpetuating during a life-time; it is a fabric of the very present and of the indicative mood.

Finally, whereas reason, like instinct, involves the whole organism, it is again on another plane. It is not purely vegetative in scope, nor narrowly vital; it is apart from the material substance of which it is a manifestation. It floods the whole organism and makes man that which is apart from all other living things.

Page 51

And yet, despite these differences, reason, we hold, is the evolutionary product of instinct. Somewhere in the past instinct revolved [evolved] into reason; at the same time, some instincts as such remained—themselves having arisen from simpler attributes. These may be traced in animals below the vertebrates to their very beginning.

Page 52

Chapter 4 - **EVOLUTION: THE NERVOUS SYSTEM OF INVERTEBRATES** – pp. 52-91

By common usage, the term, invertebrate, designates any animal without a backbone. Thus it includes the lower chordates. These were for a long time variously classified until they were placed in the same Phylum with the vertebrates because of the presence of a dorsally situated rod, the notochord, around which in vertebrates the axial skeleton is formed. The origin of the central nervous system further justifies this classification since, as in the vertebrates, the nerve-axis arises from cells of the ectoderm situated on the dorsal surface of the embryo which from a plate, converted into a tube sinks below the surface of the embryo, as we have seen is the case in the vertebrates. In no other invertebrates is the nerve axis a tube derived from dorsally located ectoderm cells.

Of the ten Phyla of multicellular organisms below the chordates, eight embrace animals which possess nervous systems. In all of them, the nerve-axis is situated not toward the dorsal, but toward the ventral body-surface. Since in those that have a blood-vascular system, the main blood vessel is dorsally located, one often speaks of neural and hemal surfaces instead of ventral and dorsal, respectively—the same terms applying for vertebrates for the dorsal and ventral surfaces, respectively. None of these invertebrates possesses a tubular nerve axis; instead the nerve cord is a solid strand of cells.

Page 53

Beginning with the arthropods, crustacea, spiders, insects, etc., usually placed next below the chordates as the eleventh Phylum, we find that in them the nerve-axis is a doubled stand with knots extending within the body along a line, the median of the ventral body-surface; that is, the strands show at intervals masses (ganglia) of nerve-cells. The largest ganglion, situated in the head above the esophagus is the brain. Behind this in the simplest types, is a ganglion for each segment of the body; but in many arthropods, some ganglia in the forward part of the body are fused. The two strands of the cord may be distinctly separate throughout the whole or a part of its length except for lateral connections whose presence gives the nerve-axis the appearance of a stepladder. The segmented worms, as the common earthworm, possess a nervous system consisting of a large ganglion in the anterior or "head" region often loosely called a brain. The ventral cord is a doubled ganglionated thread. Thus among the segmented invertebrates, arthropods and annulated worms, the nervous system is built on a similar plan, that of the arthropods being more highly developed. Sense organs are well developed in arthropods but not in these worms.

The remaining invertebrates differ from arthropods and annulated worms in lacking visible segmentation of the body. The highest of these are the molluscs (clams, snails, ink-fishes, etc.) in which the various systems show a high degree of organization; the blood vascular system comprises a heart easily demonstrated in claims and oysters, a striking structure in snails and in the squid and octopus. The nervous system, cord and ganglia, shows a progressive evolution in structure from the lowest to the highest members, the cephalopods (the squid and the octopus). In these organisms, the paired eyes are striking.

Below the molluscs one may arrange, in descending order, omitting the segmented worms, the brachiopods (claim-like animals and their allies), the echinoderms, the rotifers (wheel worms), the thread worms and the flat worms. In the mollusc-like animals, the nervous system consists of a ganglionated ring extending around the esophagus. The echinoderm nervous system, as exemplified by that in the common starfish, comprises three sets of nerve strands: a) that around the mouth with five radial extensions into the five arms (ganglion cells are found along these five extensions); b) that whose branches end at the tips of the rays in ganglia beneath the "eye spots"; and c) that, the most simple, which lies under the dorsal muscles of the central disk and the arms of the animal. A single ganglion toward dorsal aspect of the anterior region of the body is found in rotifers. The tread worms have a nerve-ring around the esophagus with six nerves extending forward and six backward. Near relations of the thread worms, the hook-headed and the arrow-worms, have ganglia. The flat worms possess two lateral sheaths enclosing fine fibers extending from a pair of ganglia anteriorly located. These sheaths, or tubes are not comparable to the tubular nerve-axis of vertebrates.

The foregoing synopsis of the nervous system in animals below the chordates, because of its brevity, may appear incomplete; but it is accurate in that it sets out the more general characteristics for each Phylum. This is sufficient for our purpose,

Page 55

namely, to trace the development of the nervous system downward by phyla—disregarding the highly interesting problem of evolution within each phylum, including variations due to habits, as parasitism, and those due to difference in habitat. Our account permits the conclusion that, from flatworms to arthropods on the one hand, and to molluscs, on the other, the nervous system gains in complexity of organization. The simplest type of nervous system is that found in the lowest multicellular animals possessing such, namely, the flatworms. In animals above these, with variations noted above, the nervous system becomes a ganglionated cord, the ganglia being segmentally disposed in the case of segmented animals. Sense organs are more and more highly developed, reaching highest stage of organization in molluscs and in arthropods.

The following scheme will aid to make clear the disposition of the nerve-axis in these invertebrates:

1. Nerve-cord with ganglia	
	Arthropods
Segmented	segmented worms
Non-segmented	Molluscs
	Echinoderms
	arrow worms
	hook-headed worms
2. Nerve-ring in anterior region of the body	
With ganglia	Molluscoidea
Without ganglia	Round worms

3. Single ganglion toward dorsal aspect of anterior part of body	Rotifers
4. Nerve-strands more or less laterally placed with one pair of ganglia	Flatworms

Page 56

The segmented worms, the arthropods and the vertebrates may be grouped together on the basis of the fact that they are segmented animals. This segmentation is externally visible in the embryonic or in adult stages or in both. There exists also in them segmentation of the nerve-axis. In the minds of some biologists, these three groups constitute an evolutionary series in the order named; some others place the worms and arthropods apart from vertebrates. Which of these views is the more tenable is here of no moment. Suffice it to say that the arthropods possess nervous systems of a high order, with the exception of the ink-fishes among the molluscs, the highest outside the chordates. In them, the spiders and insects for example, the brain is an organ of structural complexity. The instinctive behavior of these animals is one of the wonders of the animal kingdom, intriguing the naturalist and inspiring the poet.

Insect behavior however much it may have been exaggerated by workers like Fabre and writers like Maeterlinck has been subjected to most critical study by numerous zoologists. Much of insect behavior is instinctive; what of it is intelligent and reasoned, it is difficult to say. But since in the vertebrates reason overlays instinct, one can adopt the point of view that in the highest display of nervous activity in the insect world, some glimmer of intelligent behavior is present as a fore-shadowing of the more easily demonstrated higher "mental" processes in vertebrates. Accepting evolution, we are bound to accept intelligence as a product of evolution. A special creation of reasoned behavior is inconsistent with the postulate of evolution.

Page 57

Certainly, the behaviors displayed by a spider or by an insect are without doubt on a higher level than those by an earthworm. Similarly, the responses by a squid surpass those of other molluscs and of a starfish. It thus becomes plausible to correlate behavior with the structural substratum of protoplasmic organization. With the increase in complexity of exhibitions of behavior, runs an increase in structural complexity which involves all systems including the nervous. However we define the higher manifestations of nervous action, as instinct solely or as instinct plus intelligence, we are led to the conclusion that those animals with most highly differentiated nervous systems display the most advanced forms of behavior—a point which needs no extended elaboration.

All animals so far consider possess three "germ layers"—i.e., the three layers of cells that arise during the development of the embryo whence the organ-systems arise. Whatever its significance, the fact is that animals with true nervous systems, and not merely organs, have in addition to the ectoderm and the endoderm the third cell layer, the mesoderm. Other systems of organs as well appear for the first time only in animals with mesoderm. The animals without mesoderm, coelenterates and sponges, remain now to be considered.

The coelenterates include such animals as hydra and hydra-like forms, jellyfishes, corals, comb jellies, etc. Some are sessile throughout life, some free-swimming, and others have both a sessile and a free-swimming stage. These organisms possess what

Page 58

is often denominated as a "diffuse nervous system". The more frequent designation, nerve-net, I think, is preferable.

Strictly speaking, the term, system, is applied only to a combination of organs. Thus, the digestive, circulatory or respiratory system is so defined because it comprises organs closely related structurally and in the performance of functions implied in the name. A nervous system properly so-called should comprise organs, as brain and nerve cord and not only a loose net of nerve cells. Coelenterates lack a brain or a principal mass of nerve cells, a ganglion, as that found in flat-worms. That free-swimming coelenterates possess sense organs does not invalidate the argument: animals exist which have a nervous system but no sense organs. A more cogent objection to the use of the term, nervous system, in reference to coelenterates is that the structure so-called is a continuous mass of protoplasm containing many nuclei. That is, it is a syncytium like the heart-muscle in vertebrates, including man, and like other tissues found elsewhere among Metazoa. A protozoon, like Ophalina, has within its protoplasmic boundary several nuclei; one does not speak of it as a system but as a single cell. It would thus appear that the term, nervous system, does not apply to the aggregate of nerve cells found in the coelenterates.

One might ask the reason for so pedantic an insistence on the use of terms. There is first of all the interest in exact description since only by such can we attempt to formulate an explanation of that which we describe. Moreover, the rise of organ-systems in

Page 59

the course of evolution marks a step that has implications for the problem of integration. With the rise of organs in association as systems came a demand for more effective integration within each system and among the various systems of the organism. The mesoderm, lacking in no animal which has systems, furnishes a source of integration: out of it arise binding (connective) tissue, blood and blood vessels by means of which integration by chemicals is assured. Also, nervous tissue, by aggregation into organs and these in turn united to form a system, properly so-called, is another means of integration. But there is no evidence to indicate that a nervous system as such arose before other systems came into being. Granted that integration by nerves preceded that by chemicals transported by the blood—hormones and the like; this in no wise implies the earlier rise of nerves <u>as</u> <u>a</u> <u>system</u>.

The coelenterates then exhibit in response to stimuli a behavior whose morphological substratum is a diffuse nerve-net, a continuous mass of protoplasm. In higher organisms, although comparable nets be present, in some regions of the body, the nervous system otherwise is built up of integral nerve-cells, the neurons, in contact with each other by means of their cell-processes. The simplicity of behavior in coelenterates parallels the simple nerve-structure.

The foregoing discussion of the nerve-net in coelenterates, although summary, is sufficient to indicate that the component "cells", if one insists on so calling the net-work, are not <u>sensu</u> <u>stricto</u>, neurons. To name them proto-neurons in no wise serves except to add another term to the already over-burdened terminology of biology. The structure and functions of neurons are therefore best appreciated by observation and experiment not upon the coelenterate nerve-net but upon the neurons themselves, discrete, non-confluent cells in contract with each other by junction (synaptic membrane) of cell-processes. Presumably the nerve-net is a primitive nerve-structure since it is in coelenterates the chief nerve-complex and is retained in higher forms only as subordinate nerve-plexuses. But the nerve-net and the neuron-to-neuron arrangement are so different that one ought to avoid regarding them as identical.

Conduction being pre-eminently the function of the nerve-cell, any such cell, properly so-called, conducts. Since the nerve-net displays this function it has a place with the neuron as a conducting apparatus. The functional attribute resides no more in the nerve-net as a system than in the neuron-system, but rather in each case in the cellular component. The very fact that the nerve-net is not a nervous system may be taken as additive evidence that the business of conduction is carried out by a component of lower grade than system or organ. Now, according to the neurons-theory, this component is the neuron. But the nerve-net is not organized of neurons; in it conduction is carried out in a syncytium or continuum-structure and hence is diffused, not polarized as in the case of neurons. Other differences as well obtain. Nevertheless, since the nerve-net when cut into pieces retains the power of conduction, we may justify the conclusion that despite these differences, the unit of function is a stretch of nucleated cytoplasm. In this respect a tissue of neurons and the nerve-net are comparable.

Nerve-net and neurons are both derivatives of ectoderm cells; their components arise as the result of local differentiations in cells comprising the superficial layer of the embryo. Thus nerve-cell action is the action of differentiated ectoderm cells. In the early stages of development, vertebrate and invertebrate alike, these ectoderm cells arise from the upper hemisphere of the egg with the exception of those forms in which the whole superficial layer of the egg becomes ectoderm of this latter type we speak beyond. The presence of ectoderm cells is a prerequisite for nerve-cells, but in the absence of differentiation of the ectoderm, no nerve-cells arise. This is the case with the sponges.

The sponges (Porifera) constitute the lowest multicellular organisms. Like the coelenterates, they present no mesoderm. Unlike the coelenterates, they lack nerve-cells or –tissues; instead they contain some cells which exhibit more than others capacity for contraction, which some regard as muscle cells. On a strictly embryological basis one can scarcely denominate as muscle cells those not having an origin from mesoderm. If, however, one holds that these cells exemplify the notion that what counts is not embryological origin or structural characteristic but functional activity, and that therefore these cells are muscle cells, one ought logically extend the meaning of the term, muscle, to include every cell which displays contractility. In this wise,

Page 62

we would include such diverse cells as the stalked protozoan, <u>Vorticella</u>, flagellated spermatozoa, as well as motile plant cells. At the same time, proponents of this view set forth theories emphasizing most of all the strong structural characteristic of these cells. These theories we need not consider; nor do we spend time with the question, which came first, structure or function? We claim merely that the Porifera are devoid of both muscle cells and nerve cells, and the capacity of certain cells to conduct and to contract has nothing to do with their designation as nerve or muscle cells. This recalls the point often made that in coelenterates are so-called neuro-muscular cells, of ectodermal origin; these are comparable but not identical to true nerve cells. They indicate that ectodermal cells have power both to conduct and to contract. These powers are limited neither to nerve nor muscle cells; they are properties common to all living cells. Protozoa, as single cells, reveal them in high degree.

The abundance of animals and plants is one of the marvels of nature. Everywhere on land and in water this planet teems with life. No less marvellous is the reproductive power of living things. And it is not to be wondered at that men awed by this fact have become frightened by the possibility of a supersaturation by one or another species, man included, predicting dire consequences therefrom. The fear thus expressed by Malthus stimulated Wallace and Darwin, both students of the geographical distribution of animals, independently to put forward theories of a check to overpopulation.

Page 63

The abundant and ubiquitous visible forms of both animal and plant life show themselves in various sizes and shapes. From the most minute bacteria discerned only with the aid of the highest powers of the microscope to whale and giant red-wood tree, animals and plants reveal such a bewildering array of design that some biologists of eminence would have us believe that every organism, animal and plant, is a law unto itself. From such beliefs emanates the opinion that the formulation of a practicable concept for biology, comparable to that which underlies physics or chemistry, is unlikely (not possible); biology may need to content itself, it has been said, with arranging this vast horde of life-forms. The late J. W. N. Sullivan, for example, held this point of view predicting that biology is doomed to remain on that lower level in the evolution of the natural sciences concerned with classification. Sullivan, who deemed biology, as compared with physics and chemistry, a rather useless science, also expressed the view that one criterion on which he depended as a valuable diagnostic of living things, their mortality, no longer has value and is no longer valid because lines of descendants of single celled animals like <u>Paramecium</u> have been maintained in the laboratory through such long periods of time that these forms may be considered as immortal. Similar claims have been made for tissues of warm-blooded animals. Every reader has doubtless seen the annual announcement in the daily press that cells isolated from an embryo chick were still kept alive. But these laboratory experiences, undoubtedly interesting, have little bearing on the question of immortality. As generally employed, immortality is a term that connotes the indefinite maintenance of individuality; and individuality

Page 64

being fleeting—the more fleeting, the more individual—the conception of immortality is inconsistent with that of individuality. Impermanence and dynamism are cardinal attributes of the life-state, and death is foretold by life, cradled with it, so to say, at its very beginning. Mortality we may designate as a factor in producing the multicolored variegated pattern of the living world.

Livingness can be defined other than in this negative manner, in terms other than the interdict of death; living things are set off positively from non-living, the animate is separated from the inanimate, by a degree of organization encountered nowhere else in the natural world. An opinion that the science of living things has no foundation comparable to the clearly defined conception underlying physics or chemistry thus overlooks organization of substance peculiar to animals and plants. All natural phenomena are emergent of structure, of organization, more or less easily comprehended depending upon its amenability to analysis. The peculiar organization of matter, named the life-state, defies analysis insofar as this destroys life. Hence, the resolution of the problem, what is life?, is to be sought first of all by way of the most complete recognition of an organization, superlative to that of inanimate systems, whence are displayed vital manifestations.

Now, generally, one speaks of this organization as cellular. The cell is usually considered the unit of living substance, a mass of protoplasm comprising a kernel, the nucleus, and a surrounding body of stuff, the cytoplasm. For the great majority of living things, all multicellular organism being made up of cells, this statement holds. But there exist living forms, as certain

Page 65

bacteria and other lowly plants, in which nuclei as such are lacking. In the strict sense, then, these are not cells, even if they possess in the form of dispersed granules the material found in the discrete nuclei of true cells. Moreover, many simple organisms consist of a mass of cytoplasm in which is imbedded not one but several nuclei—a condition met with in some tissues of highest animals and similar to the nerve-net as already described. These a-nuclear and multinuclear forms of protoplasm, being directly comparable to the true cell, indicate that the presence of a discrete and single nucleus is not a common property of protoplasm, the structural unit of living things. Nuclearity, <u>per se</u>, is not an all-pervading diagnostic of that organization which sets apart living from non-living substance.

No matter how impressive the variety of living forms, we find always that each has specific size and shape; for each life-form there are limits beyond which it never passes. Hence, instead of being an obstacle to the formulation of a common concept that holds for all living things, the multitudinous variety of the forms of life challenges us to find in the limits of each living thing an expression of the specificity of its protoplasm which sets these limits. Size and shape as specific attributes of the living thing are undoubtedly intrinsic to the protoplasmic organization. But since this organization does not always comprise nucleus and cytoplasm, we need another character common to all living protoplasmic systems. This is found in the make-up of the cytoplasm into an inner core and a peripheral layer; no protoplasmic system exists that does not exhibit this spatial and functional differentiation of the cytoplasm. I do not maintain that this peripheral

Page 66

layer alone determines size and form specific for a given living protoplasmic mass. Nor do I say that only living things have definite size, limited in space. What I insist upon is that given the fact that size and form are specific for each life-thing, the peripheral layer of the protoplasmic system plays a role in limiting the space-pattern of the protoplasm. Since every protoplasmic system has specific limits, a limiting surface is its sine qua non.

Ectoplasmic structure has long been used for classifying Protozoa. On the basis of their means of locomotion, Protozoa are sub-divided into four classes: 1) forms that move by means of pseudopodia—i.e., by the protrusion of more or less delicate processes; Amoeba and allied forms; 2) Forms that move by flagella or whip-like processes; 3) Forms that move by cilia; Paramecium, the slipper-animalcule, and allies; and 4) Forms which in the chief phase of the life-history show no means of locomotion, are quiescent and spore-like; the malaria parasite and its allies. The means of locomotion are all ectoplasmic formations. Protozoa are well known for their strongly revealed ectoplasmic structure.

Moreover, by observation and experiment it can be shown that in animal egg-cells this limiting surface, the ectoplasm, has its own specific structure superposed upon the specificity inherent in the cytoplasm.

Without doubt every animal egg exhibits this differentiation of the outer rim of cytoplasm from the inner; it has been described for eggs of animals representing all phyla of animals that have eggs. Indeed, so striking did this appear to some earlier workers that they contended that only this outer region of the egg be alive. Although we may speak in general terms of a "conventional

Page 67

design" of ectoplasm, as composed of radial threads that are prolongations of the living cytoplasm and a thin membrane covering the outer ends of the treads, the ectoplasm of even closely related species of eggs shows differences in structure.

Here questions concerning the structure and the properties of this surface assert themselves, demanding answers. Although a region found in every protoplasmic system it may be merely an envelope passively maintaining the integrity of the life-state, a region, comparable to that spoken of as "the dead space" in chemical reactions in capillary dimensions, where the life-processes set up a barrier of inertia against the medium external to the living system. On the other hand, since it portrays itself as being differentiated from the inner core of the protoplasmic mass, the postulate lies close at hand that, being living, it has a role in vital behavior other than that of restricting size to that specific for the system. Elsewhere I have presented an array of evidence in support of my theory that vital manifestations express themselves in the behavior of this peripheral region of the protoplasmic system. For the present purpose a brief review of this evidence suffices.

First, let me justify the statement that the protoplasmic surface is structurally a _sine qua non_ of life. Thus expressed, the statement seems ludicrous—how could matter lack a bounding surface! And yet there are biologists who deny its presence. It is not strange, therefore, that the existence of this surface, as a living and differentiated structure, has also been denied and when not has been allowed only for some protoplasmic systems, being held for these in some instances as artefact, a product of the observer's technique.

Page 68

Very well do I appreciate the fact that this failure on the part of biologists generally to recognize the ectoplasm as a living region common to all protoplasmic systems, a failure easily understood as an outcome of the undue value placed upon the rôle of the nucleus, a structure not always present in protoplasm, may appear to many readers as creating a formidable obstacle to the acceptance of the thesis developed in the present essay: one ought not to hazard the elaboration of a theory grounded on a not generally recognized basis. Nevertheless, in the present instance, the basis, ectoplasm as structure common to protoplasmic systems, stands out so clearly that reproach, if such there be one, is not to be ledged against those who recognize it. Its visible structure and easily apprehended activity are facts, however much overlooked.

The strongly expressed differentiation of the cytoplasm of uni-cellular animals, the Protozoa, into endoplasm and ectoplasm has long been familiar to students of this group of organisms. The differences, in optical properties, in consistency and in the distribution of granules, between the two regions have been amply described. Probably no biologist would deny the existence of ecto-endoplasmic differentiation among the Protozoa.

These minute animals, most of them are visible only under the microscope, exhibit marvelous designs in ectoplasmic structure; their protoplasmic surfaces give rise to modifications of amazing diversity equaled nowhere else in the animal kingdom. On the basis of differences in modification of ectoplasmic structure, Protozoa are sub-divided into four classes, the means of locomotion furnishing a simple diagnostic for the classification. Certain forms, including Amoeba and its allies, move by means of protruding more or less delicate processes, or pseudopods; some, the Flagellates, move by lashing back and forth whip-like extensions of the ectoplasm; those of the class to which belongs Paramecium, a favorite object of study, move by means of cilia, processes that are shorter and more numerous than those of the flagellates; the fourth class, the Spore-animals, including, for example, the malaria parasite, are spores during one phase of the life-history—hence the name. Thus, among the Protozoa, not only is ecto-endoplasmic differentiation well known, but also are varieties of ectoplasmic designs easily demonstrated; for in addition to modifications for locomotion, others have been abundantly described as, for example, for protection and support, for capture of food, for throwing off fecal matter and for the elimination of the end-products of metabolism.

Far less stark than, but nonetheless as visible and tangible as, the ectoplasm of a protozoon is that of the animal egg. Both protozoon and egg are single cells; but, whereas in the former the cycle of life is completed within its narrow confines, in the latter there is no such shortened circuit. Singleness for the egg is but a pause before that briefly enduring moment when comes that shock which elevates the egg to the level of multiplex organization and which failing, death intervenes. As we shall see, this upward thrust is achieved by events localized in the ectoplasm: the surface of the dormant egg is shattered and then remolded into a number of surfaces to make the multicellular organism.

It may be that the temporary state of singleness accounts for this less sharply discriminated ecto-endoplasmic differentiation in eggs. Certainly, it is true for some species, when in this pause, that one can delimit the ectoplasm only by very careful observation; also in such, the boundary comes out sharply with the use of simple and innocuous experimental means. That ecto-endoplasmic differentiation

Page 70

is limited to only some half dozen species of eggs, as certain famous biologists assert, is thus refuted; it is furthermore gainsaid by careful observers of repute whose clear descriptions, in treatises reckoned among the greatest contributions to embryology, may be found by anyone.

Once the egg has left the single-cell stage, during the period marked by successive sub-divisions whereby it assumes the multicellular conditions, even a novice can now readily observe the ectoplasm as differentiated from endoplasm, the inner region of the cytoplasm. With a modicum of capacity for microscopic observation, he can learn that the whole is not a mere loose aggregation, but is a mass of cells adhering by virtue of ectoplasmic prolongations.

This differentiation of the rim of cytoplasm from its core, has been described for eggs representing all the phyla of egg-bearing animals, from sponges to mammals. Indeed, so striking did it appear to some earlier workers that they contended that only the outer region of the egg is alive. Although we may in general terms speak of a conventional design of the ectoplasm of egg-cells—i.e., a structure made up of radial threads, prolongations of the living cytoplasm and a thin covering membrane formed by the fusion of the out tips of the prolongations—it differs even in closely related species of eggs. Also, in a named species of egg, it varies structurally with stages of development. When eggs are "sick" or dying, they exhibit alterations in ectoplasmic structure, a fact against which the French zoologist, Giard, warned because, in his time, workers had mistaken these abnormal changes for normal. So do some today make the same mistakes.

Page 71

Capable of being traced through all stages of development, ecto-endoplasmic differentiation is found also in cells of adult animals. The ectoplasm in such tissue-cells, including those of man, has been abundantly described by the older microscopists; it shows up strikingly in cells of the skin, of the retina of the eye, in cells of the gut, etc. as in egg-cells, the ectoplasm of tissue-cells presents numerous filaments which interlace with those of contiguous cells. In view of the fact that these projections, often called intercellular bridges, can be beautifully demonstrated on living cells—many of the early descriptions were so obtained—it ought not be insisted upon, as by some, that these formations are artifacts.

An elegant portrayal of ectoplasm in living tissue-cells is that given by the cultivation of tissues removed from the body of an animal. By this, the tissue-culture method, first employed by Harrison in his work on embryonic nerve-cells, not only the structure, but also the behavior of the surface-located cytoplasm can be easily ascertained. The extensive literature on tissue-culture abounds with descriptions of this structure and its behavior.

Thus, the ectoplasm is a structural constant of protoplasmic organization. This statement holds not only for animals, but also for plants. Bacteria possess well-marked ectoplasm, as do other plants. In many of the latter, only the peripheral layer of the cell-contents is living. Protoplasmic systems lack or contain a nucleus; but they are never without a living boundary layer differentiated from the remainder of the system.

Evidence, proof, one could say, insofar as any problem in nature can be proved, can be adduced concerning the great rôle played

Page 72

by the ectoplasm in vital behavior. By its displays of activity we grasp the manifestations of the life-organization that otherwise escape us. The protoplasmic system, let me repeat, is the unit of the state of being alive; then it ought not to be envisaged as a mere aggregation of separately acting particles or of regions even. For, to argue for the aggrandizement of the ectoplasm in vital functions could be as fatal as to assign hegemony in the life-state to the nucleus—i.e., in protoplasmic systems that contain nuclei. Although we know that a cell cannot maintain its life after removal of the nucleus, and that the nucleus plays some part in heredity, we have not a single datum of one definite physiological action on its part. This ignorance has not hindered the setting up of claims not merely for the nucleus but even for a separate particle, the gene, of a nuclear structure, the chromosome, as the end-all and be-all of life. Despite, or because of, the fact that the behavior of the ectoplasm, on the other hand, is visibly displayed, easily tangible and readily apprehended, it need not be exaggerated.

Although the evidence, by no means scant, of the behavior of the ectoplasm, is not available for all animal cells, it is nevertheless as impressive as that of its structure. One may safely assume that more investigations by experiment on this hitherto so greatly neglected region of the living protoplasmic system will, by embracing more species, widen our knowledge of its physiology. Facts at hand warrant the drawing of definite conclusions concerning the rôle of the ectoplasm in expressing vital manifestations.

The vital activities of any Protozoon, being a single cell, obviously must be interpreted as those of such, of one continuous

Page 73

mass of confluent cytoplasm containing no, or one or more, nuclei. Each is a single complete unit, completely independent of any other, in, on, for and by which all its vital reactions are carried out. To speak of organs in these single cells is a gross error, to name them little organs (organelles) is just as bad: an organ, big or little, in a living form, is an assembly of tissues, themselves comprising cells. Protozoa are often tremendously complex, but that this complexity demands structure identical to the multicellular condition is a gratuitous assumption not admissible by facts. The physiology of Protozoa is therefore the physiology of single cells. For the investigation of physiological processes in multicellular organisms, a most important task is to locate the processes in the various systems, organs and tissues. Years elapsed after Lavoisier's epoch-making researches on animal combustion, the foundation-work of modern physiology, before the seat of respiration was located. We still name efferent blood vessels of vertebrates, arteries, because the older anatomists regarded them as carriers of air. Physiological study of the Protozoa entails no such duty; we know that the activities displayed are those of a single protoplasmic structure, of a unit. Without violation of this unity we endeavor to locate, in the more water, ever changing, ever circulating inner and in the less watery and as changing mobile outer region of the cytoplasm, those activities that maintain the integrity of the Protozoon as a specific living thing.

Now it cannot be that the digestion of food has place except within the interior of these animalcules. Unlike digestion in the human mouth or gut, accomplished external to the cells, protozoan digestion is intra-cellular. In an Amoeba or many another Protozoon, one can observe the digestive process as occurring in the endo-plasm—especially in Paramecium with its digestive—gastric (!)—"vacuoles" is this observation facile. So too the process of assimilation, the building up from the raw food-stuff of new protoplasm, is a function of the internal cytoplasmic menstruum. If, in the transformation, the protoplasm increases, growth results. Growth, as one says, is in all organisms by intussusceptions rather than by accretion, the mode for inanimate systems—a difference so great that the term, growth, inadequately defines one mode if it properly defines the other mode. Growth, as the addition of new living protoplasm, comes about through the activity of the endoplasm; cells grow from within.

By no other path except across the boundary layer of cytoplasm does food reach the cell's interior. This is truly a banal statement, for no substance, solid, solution or gaseous, can otherwise penetrate the cell. Hence, for digestion and assimilation, its most vegetative functions, the Protozoon depends upon the activity of its surface structure. The surface may be only in a temporary state of modification, as is the case of that of <u>Amoeba</u> when it thrusts forth pseudopods, or the modifications may be permanent, as with the "mouth" and "gullet" of a <u>Paramecium</u>: the peripheralized cytoplasm is the means for the capture and the ingestion of good. Even in cells that require no solid food and whose nutrients are fluid or liquid, the activity of the living outer layer of the protoplasmic mass conditions the satisfaction of the requirements. Antecedent to endoplasmic reactions of assimilation, of repair for maintenance, and of accumulation as growth, these reactions devolve upon the ectoplasm.

Page 75

The products of wear and tear as well as the end-products of energy-changes, together with undigested residues, are eliminated by means of the ectoplasm or of its transient or permanent modification thereof. It often builds intricate and exquisitely patterned shells, armors, tests, etc., either in whole or in part of itself, for support or for protection; in some species it elaborates bodies for defense.

Ingestion, necessary antecedent to digestion, clearly implies an underlying response of the organism to its surroundings; so too egestion and excretion which, whether by leakage or by more active expulsion, take place in consonance to pressure conditions within and without the living protoplasm. The dependence upon the external medium of the ectoplasm for the elaboration of supporting or protecting structures need not be argued: many of these, if not formed directly of solid particles stuck to the cell-surface, are indirectly produced from constituents present in the environment. Highly irritable, keyed even to slightly fluctuating changes in its normal surroundings, the organism displays its specific behavior in response. It reacts to stimuli received and transmits the impulses as effects of stimulation. The transmitted impulse blazes along the ectoplasm by virtue of its location; the ectoplasm is, affected by the external milieu before the inner more watery menstruum, which is laden with food in various stages of digestion and with effete matter to be discharged. Even after disruption of the cell, the transmitting power of the ectoplasm persists, whereas the exuded endoplasm, disintegrating, goes into solution.

Not only transmission (or conduction) but also contraction is a property of the ectoplasm. The physical make up of the endoplasm

Page 76

is as much against as that of the ectoplasm is for contractile power; the elasticity of the latter is superior to that of the former. Structures accentuating contractility, as the stem of <u>Vorticella</u>, a ciliated protozoon roughly resembling a wineglass with a cover, are ectoplasmic.

Attending both conduction and contraction, are the energy-changes, and the exchange of gases, such as oxygen for carbon dioxide, that make up respiration, without which life does not maintain itself. Throughout the entire protoplasmic system, oxidation takes place, for it underlies all physiological processes. That only the nucleus is involved is now an abandoned theory. The taking up of oxygen is an ectoplasmic property; the toxic carbon dioxide is got rid of by the protoplasmic surface.

These activities along with those underlying the transformation of food into living substance, subserve the maintenance of the Protozoon as self; somewhere enshrouded by them is that mystery of life, its capacity for self-regulation, that so sharply distinguishes the animate from the inanimate system. Together they hold the light of life that strikes for a moment a luminous spot and is then gone. But this light we do not know, may never know. Blind-worshippers, we, sitting in dreams or speculation; and we augur the multitudinous flames, no less hidden, that glow along the ladder of evolution.

This maintenance, of self, is life seen from one angle or at short range; however long, as we measure time, a member of any species of plant or animal may persist, the life of an individual is fleeting; among organisms as bacteria it may endure for minutes only. Intangible though the life of the individual may be, it is less

Page 77

so than that thrust we name the perpetuation of the species. What we designate as reproduction is, if a teleological expression be permitted, the aim and goal, the purpose even, of life. To reproduction the selfishly engaged activities are subordinated. Life exists only to be perpetuated.

What is meant by the term, reproduction, as used in biology, is to be dissociated from all other meanings of the word. Indeed, no connotation, even the slightest, ought be carried over from any of these to the definition of the biological process. The copy in the same medium of an old master, presumably inferior to the original; of a Paris gown made by hands other than those of the creator—must certainly be inferior even with the same stuffs. Likewise, a procession of automobiles, of cash registers or of card indexes from the same manufacturer; an elegant experiment or a nice observation in science, exactly repeated; the stimulation by a teacher of his students, who gloriously continue his work—by these and many other examples, trains of thought are ignited by the word, reproduction. The reproduction of animals and plants is something other; there is really no happening like it elsewhere in nature.

What is reproduced by a living organism is only a conformation to a standard set by man and learned to be true through experience. An _Amoeba_ or a _Paramecium_ reproduces over and over again, not itself but an entity that meets the specifications of this standard; it matches this picture that our minds form of _Amoebaness_ or _Parameciumness_. For, mark you, even when a Protozoon "reproduces" by sub-division into two identical equal portions, if this ever occurs—and so-called equal division is by no means invariable, there being innumerable cases of multiple and of unequal division— what emerges now is a new individual whose identity to the "parent" is proportional to its content. And from the moment that division is consummated in the organism, new relations are established for each piece due to its now diminished quantity, to say nothing of the altering of qualitative conditions by which the division was called forth. Hence, even the "fractional identity" of the "offspring" with "parent" is limited, so limited that I wonder if we are permitted to speak of identity as being at all retained or of individuality as being contained.

Protozoan reproduction entails the establishment of the new products of division finally on that level which characterizes the parent. If, for example, by repeated divisions the separated "daughter-cells successively diminish in size, the organism would, failing some mode of reconstitution, disappear; this would certainly not be a reproduction. Then, following the setting up of new individuals by division, growth intervenes. Reproduction thus, from this point of view, only begins with division; its true end-point comes when the daughter cells attain the parental level. Before this, not only size but a host of attendant processes make it a being something other than the parent was at the time of division. Reproduction looked at thus is not a suddenly-occurring event; it is a process that spreads over time, which varies according to species, a building-up process that carries the offspring to the parental level. We may speak of division only as the <u>initiation</u> of reproduction.

Some intra-protoplasmic state, as culmination of the processes for maintaining self, engenders division as the means of reproduction; then the self disappears. There is no continuity of the individual life but only a maintenance on a wider plane, that of the type. There neither is nor can be through reproduction immortality; these terms are self-contradictory. Immortality, be it of Protozoa,

Page 79

of cells cultured outside of the animal's body, or of the germ cells of multicellular animals, is a stimulating idea; but alas! in the cold light of science it vanishes, as a dream, however beautiful, fades with waking in the luminous dawn.

How does it happen that a protozoon reaches that state where reproduction becomes imperative? In other words, what qualitatively or quantitatively underlies cell-division? On this point our ignorance is profound, owing less perhaps to hazy definition of the term, reproduction, than to an incomplete knowledge of the phenomenon in question.

Usually biologists describe what they call the mechanism of cell-division as centering in mitosis, the elaborate maneuver by which the nucleus divides. But this cannot be fundamental since some protoplasmic systems, as we have seen, lack a discrete nucleus. Moreover, in cells, defined as protoplasmic systems possessed by one or more nuclei, the nuclei do not always divide by mitosis; in many cells, both of protozoa and multicellular organisms, nuclei often divided by a-mitosis. Further, the nuclei divide by either mode with or without concurrent division of the cell-mass. Finally, division of the protoplast not always being equal, as is often stated even by specialists of this subject, cytoplasm may be unequally distributed to one or the other of the resulting components. Hence, the equivalence of the nuclei in dividing cells, so generally agreed upon by students of the chromosomes, is often opposed by marked disparity in amounts of cytoplasm. For these and other reasons that I have elsewhere set forth, we cannot causally relate division of the cell to division of the nucleus. Whatever the cause of cell division, it lies outside the nucleus.

Page 80

Here is a protozoan mass of protoplasm consisting of cytoplasm and its contained one or more discrete nuclei which by division gives rise to two or more complete animals. In the simplest cases, division into two approximately equal cells, one can establish beyond question that the two components have the same structural characters. There is therefore a reconstitution whereby arise duplicated structures. Division is never accomplished by means of a pouring out, as from a container, of the more watery internal stuff of the cell which then sets in the form characteristic of the species. Rather, the cell is cut into two by a constricting ring which begins at the surface and progresses inward. However complicated the mode of cell-division in whatever animal or plant is, it is so accomplished that the surface of the cell is partitioned off to the daughter cells. There is no cell-division otherwise.

When the protoplasmic mass of an animal egg divides, it is the mobile ectoplasm that brings about the separation and the reconstitution. Thus, division in the egg-cell is similar to that in protozoa. Other processes in egg-cells are related to and explained by the activity of the cell-surface.

Experimental evidence of various kinds indicates that fertilization of the animal egg is a reaction between the egg's surface and the spermatozoon. The ectoplasm, necessary for fertilization, is the seat of chemical changes which underlie visible form-changes, characteristic for the species of egg and never again repeated in its life-history. Similarly, in the initiation of the egg's development by an experimental means which replaces the spermatozoon—in experimental parthenogenesis, that is—the visible ectoplasmic changes are prognostic of the quality of the development that follows. During the development of the animal egg, the cells continue to adhere, although they come into changed spatial relation in a more or less time-ordered fashion is constitute the fundaments of what are the primordial embryonic tissues whence arise the animal's organs. This connection is insured by the ectoplasmic filaments whose differential activity is the chief cause in shifting the cells to new positions at just the right moment. Thus, from fertilization or the experimental induction of development (experimental parthenogenesis) onward, the filar activity of the surface-located cytoplasm under-lies the orderly integration of the cells, the striking phenomenon in the transformation of egg into embryo.

Unlike the majority of spermatozoa, which through the possession of an ectoplasmic filament, the tail or flagellum, have locomotor power, animal eggs, as a class, are non-motile, although some show amoeboid movement. Others reveal strong spontaneous contractions and all animal eggs subsequent to fertilization exhibit contractility in some degree. This property is ectoplasmic.

The ectoplasm of eggs, like that of the Protozoa, portrays capacity to conduct, as can be shown especially well at the time of fertilization. Then a wave of conduction sweeps over the egg-surface, in a definite manner characteristic for each species of egg, beginning at the point at which the spermatozoon enters. For the endoplasm no such power can be demonstrated since it does not extend to spermatozoa.

When contraction and conduction are displayed in consequence to fertilization, the ectoplasm consume oxygen and gives off carbon-dioxide in greater amounts than at any other moment in the life-history. To be sure, the whole of the living protoplast respires; but the evidence permits the conclusion that the ectoplasm of the egg

Page 82

is the seat of greatest gas-exchange.

Water, quantitatively the most important compound in protoplasm, can be demonstrated in living, viable egg, as in other cells, to be present, both normally and under wholly innocuous experimental conditions, in discrete drops. Obviously, water, or any other sub-stance, can enter cells only by way of their surfaces. Moreover, since during periods of rhythmical cellular activity, as has been proved, the peripherally located cytoplasm undergoes concomitant changes in structure and function, it is easy to assume that the rhythmical variation in a cell's water-level depends upon ectoplasmic behavior; observation and experiment make of this assumption a scientific fact.

Animal eggs, being single cells, express as do Protozoa the behavior of such. In like manner, individual cells of multicellular organisms display activities either similar, and even identical, or strictly comparable, to those of Protozoa and of eggs. These activities include manifestations located at the cell-surface, properties common to ectoplasm wherever found. In addition, as one would expect, are ectoplasmic expression peculiar to Protozoa and to eggs and to cells of the adult state of higher animals, for these cells constitute three distinct categories of cellular organization. Singularity of state, which in Protozoa is beginning and end of the life-cycle, in eggs only beginning; and in multicellular organism it is a single term with little meaning except in relation to others in the final complex.

Unity of behavior in the unicellular organism is circumscribed by the boundary of singleness; in the multicellular organism, by interdependence established by integration of cell with cell. This is the first and perhaps greatest function of the ectoplasm in cells of

Page 83

complex animals. The harmonious coordination of functions, a marvel of nice adjustment, in the most complex multicellular animal makes a unit of myriads of single cells. To explain this unitary action, it is not necessary to invoke any semi-mystical principle as the organism-as-a-whole, thereby abandoning the ground-postulate of the structure and function of living substance, i.e., the organization of protoplasm set up within discrete boundaries. Rather, it is our duty to seek to explain why protoplasm is so confined and how in the case of multicellular units the constituent protoplasmic compartments comparable to those capable of existing as individual organisms, lose their independent freedom of action; how, instead of being a loose agglomerate, they become a knitted pattern whence arise a display of coordination so close that the organism behaves as a single entity. Study of the structure and the behavior of the living protoplasmic boundary goes far to answer these questions. A mass of living protoplasm, a cell for example, exhibits its tangible expressions of livingness at its surface; directly or indirectly, vital phenomena have location here. For protozoa and for eggs this is a banal statement. In addition to the fact that matter in nature is never without limit, is this fact: the ectoplasm, not merely by virtue of its location, but more in view of the definite evidence showing its precise functions, is the seat of just those activities that set living things apart from the non-living. Among cells in a multi-cellular organism, common sense would tell us, that connections are established by surfaces; observation and experiment confirm this. The ectoplasmic activity of first and greatest importance for the unity of a multicellular organism is that of establishing and maintaining intercellular integration.

Page 84

Now, in multicellular animals, integration by structure—I leave aside chemical integration, i.e., by internal secretions not always capable of being demonstrated as present in animals—may be classified as: that by cell to cell in the same tissue, that by nerves, and that by connective or binding tissue. Sponges, the lowest forms of multicellular animals, possess integration by cell to cell only. Although coelenterates have nerve-cells, nervous integration for most of them is less important than integration by cell to cell. The normal development of any animal egg depends in large parts upon the spatial and temporal disposition of the blastomeres, and this in turn upon the integrity and differential activities of the connections of blastomere to blastomere. Thus, cell to cell integration is not only the oldest mode of integration in the evolution of the animal kingdom; it is also the first to arise in the development of any egg.

Nervous integration, feebly developed, appears first in the coelenterates; steadily progressing in development, it reaches its highest form in the vertebrates. A striking feature of the development of vertebrates is the early rise and precocious growth of the central nervous system, a vast tract of cells linked to each other by their connections. But although early laid down, the nervous system arises later than the blastomeres whence it originates. Likewise, nervous integratin is of later origin in the development of a given species of egg, just as it is in the evolution of the animal kingdom.

Integration by connective tissues, both in adult animals and in the course of a given egg's development, also occurs subsequent to the appearance of intercellular connections. A derivative of mesoderm, it occurs obviously only in animals and in eggs that

Page 85

possess mesoderm. In lieu of mesoderm, coelenterates have a jelly-like substance that serves somewhat the functions of connective tissue. Connective tissue integration may be designated as passive or sustaining.

What is common to these modes of integration? After all, the timing of the rise, whether early or late, in the evolution either of animals or a single species of egg, of a named mode of integration may not at all indicate its significance. The passive binding by connective tissue, even including cartilage and bone, would scarcely be regarded by most readers as being as significant as the coordination achieved by nerves. However primitive may be connections between adjacent cells in a tissue, this means of integration is certainly on a lower level than that by nerves. The emphasis here is not altogether on the priority in appearance of the cell to cell integration; it is also on a point too generally overlooked by biologists, namely, that all these modes of integrations have a common basis, that although they stand at different levels, they exercise fundamentally the same action but to different degree. Simply stated, this point is that all three means of integration are ectoplasmic; they differ in mode in that they integrate like or unlike cells.

The existence of intercellular bridges, ectoplasmic prolongations of the living protoplast, among cells both of developing eggs and of adult tissue, is a demonstrated fact beyond question. There are biologists who deny this—so much the worse for them! Beyond dispute also is the display of vital functions by these connections, one of which, less definite but none the less real, is the coordination of the individual cells. Break these contractual threads, and what was a mass of cells behaving in harmony as a unit

Page 86

becomes a collection of independently acting single entities.

Consider a mono-embryonic egg, that is, one which after fertilization normally gives rise to a single adult—there being poly-embryonic eggs, as certain insects', that normally develop into more than a thousand embryos. Normal mono-embryogenesis ensures only if the cells resulting from the successive sub-division, cleavage, of the egg remain in normal contact. If, however, the partitioned chambers be separated, one of two possible results, depending upon the species of egg, is obtained: defective, or partial, embryos, or complete, though dwarf, embryos. In the latter group of species, the cleavage-cells, by retaining contact through the un-impaired integrity of the interlacing ectoplasmic filaments, suppress any inherent capacity for multiple embryo-production. In the former group of eggs, this capacity is suppressed before cleavage; for from them, as from all eggs amenable to the experiment, if sub-divided before fertilization, one perfect though dwarf embryo is developed for each fragment fertilized. The puzzle that nature has given us and which we have yet to solve, that is, how out of a microscopic speck of protoplasm whose diameter is around 14/2500 inch a whale or a human being can develop, is complex enough. To this wonder of development from a single egg is added that of the suppression of multiple embryo-production. And this suppression, the restriction of many to one, is performed, guaranteed, by the cell-boundary.

No less do the ectoplasmic connections exercise restraint upon the inherent independent actions of cells in tissues. Such cells, if removed from the living organism and culture _in_ _vitro_—the method of tissue-culture referred to above—manifest a behavior—that is the result of their having escaped from normal contact with their neighbors; they give play to larger freedom of movement, exhibit contractility to greater degree, and display more their capacity to multiply. Even without the technique of tissue-culture, too often needlessly elaborated, one can by simple direct observation on separated tissue-cells demonstrate that their behavior changes upon isolation. One such simple demonstration is the following:

A strip of cells, each of which bears a delicate brush-work of threads, cilia, projecting from one surface, is removed from an animal—preferably cold-blooded for its tissues are more easily kept viable and normal than those from a warm-blooded—and mounted, in its own body-fluid or that of a nutrient solution having the same salt concentration as that of the animal's blood or body-fluid, and viewed under low power of a microscope. The cilia can now be observed to beat in an orderly progressive manner, beginning with those in one cell and passing to the next. Thus, wave after wave of bending and return of the cilia sweeps along the length of the strip of cells. Their successive fall and rise make one think of the movement of blades of wheat yellowing in the August sun under the gentle pressure of a wind sweeping over the field. But if the cells be separated, the rhythm is lost, and the cilia beating independently and without coordination now that cell-to-cell contact is destroyed. That the successive fall and return of cilia that course progressively along the intact strip of cells is due to surface contact is illustrated by another example.

Normally flagellated animal spermatozoa move independently and a-synchronously; a hundred such of a marine animal suspended in sea-water and viewed under the microscope swim helter-skelter, propelled by the lashing of their tenuous tails or flagella. But if, as the embryologist, F. R. Lillie, has shown, they are made to form a ball, by the introduction into the suspension of a substance bringing about the temporary adhesion of their surfaces at the anterior tips, leaving free the tails at the opposite ends, the tails fall and rise in succession in beautiful rhythm. The cluster then breaks up and the a-synchronous lashing of the tails returns.

Integration by nerves also is due to ectoplasmic processes among which are the most extensive known for nerve fibers, shown by Ross G. Harrison to be ectoplasmic filaments, may in man attain a length of one meter or more. Likewise, the passive integration exercised by connective tissue is ectoplasmic; according to the late Franklin P. Mall, connective tissue fibres are cast-off ectoplasm. The no longer living ectoplasmic filaments, the fibers of connective tissue, exercise only passive integration. Moreover, integration, be it of like or unlike cells, is at basis the same since in living tissues, both nervous and non-nervous alike, it is due to the capacity of the cell-surface to transmit impulses. Transmission, or conduction, is a property common to tissue cells. As

92

we have seen, the surface of the protozoan and of the egg cell has this same capacity; hence conduction is a general property of the living ectoplasm. Thus, all forms of integration by structure are to be attributed to the peripheral cytoplasm.

That this function of the cell-surface embraces like or unlike cells does not imply great differences in the mode of integration. As a matter of fact, the difference is not absolute. Take the nervous system, for example, by whose cells all parts of the organism are brought en rapport. Within it, nerve cells make contact with other nerve cells; integration is here therefore the primitive mode of cell-to-cell integration, the means being finer and more exquisite—by the rich and easily visible neuron-processes.

In other respects the peripheralized cytoplasm resembles the limiting surface of Protozoon and of egg: it has contractile power, is capable of ingesting material, and plays a rôle in water and gas exchange and in the elimination of waste. Contractility whilst pre-eminently the function of muscle-cells is by no means limited to them and to the wandering, amoeboid, white blood cells. Every living viable tissue-cell in the intact organism has power to contract, although this may be feeble. Contraction by various types of cells, including nerve, is easily demonstrable in tissue culture. Indeed, the growing tip of the nerve-cell, its ectoplasmic filament destined to become the nerve-fibre, establishes contact through its contractile power; this is likewise true of regenerating nerve-fibres in the organism. Just as nerve-cells emphasize the property of conduction common to all living protoplasmic systems, so in muscle-cells, themselves endowed with the power to conduct, contraction reaches the peak of excellence.

Cells of all organs need food, water and oxygen; they get rid of waste products. These substances cross the cell surface, which by modifications in structure facilitates their passage. It has long been known that the "eating" cells of the blood and of the liver in vertebrates, the wandering and the fixed tissue phagocytes, respective, are capable of ingesting particles. The former, as Metchnikoff's elegant researches showed, engulf bacteria, the latter remove from the blood the iron freed from red blood corpuscles that daily break down in the liver. This phagocytosis has recently been "discovered" anew by students of tissue culture and given a new name—to make the "discovery" appear more important. Cells of the gut likewise ingest solid particles. Phagocytic power is undoubtedly common to all living cells, a fact which renders superfluous adherence to any one of the various conflicting theories of semi-permeability, sufficient in number to satisfy almost any taste, with all their glaring inconsistencies both as to logic and common sense knowledge of biology, and according to which cells cannot take in nutrients. But common sense tells us that they do, since they live and often grow. The ingestion of substances in solution or in minute particles is due to the richly filamentous, ever active superficial cytoplasm. These filaments behave as the pseudopods of an Amoeba. Nowhere in the animal kingdom does there exist a perfectly spherical cell or one with a regular, smooth, quiescent surface. Thus, the ideal cellular state demanded by the "semi-permeabilists" and the "surface-tensionists" is pure phantasy. Books like the admirably written erudite *On Growth and Form* (1917) by Professor D'Arcy Thompson, so pleasing to read, which reduces many life-processes to a matter of the mathematics of simple geometrical designs, as the sphere, etc., belong in a category of biological literature comparable to that that includes Plato's *Republic* or More's *Utopia*.

What is said of food applies to water and to oxygen. To enter the cell, they must cross its boundary. The extremely fine ectoplasmic filaments are admirably disposed to facilitate the taking up of water and of gases. It is obvious that whatever leaves a cell leaves it by way of the surface.

Page 91

Be it protozoon, egg or cell of a multicellular organism, ectoplasm, the peripherically located cytoplasm, is always present as a structure differentiated from the inner cytoplasm. Hence, outside of the nucleus, the living stuff shows ecto-endoplasmic differentiation. In non-nuclear protoplasmic units, as bacteria, the same cytoplasmic differentiation prevails. Therefore, the ectoplasm is a _sine qua non_ of living substance, the discrete nucleus not.

Protozoa are units of fixed, whereas eggs are of transient, and tissue-cells of virtual, singleness. Correlated with these different values of singleness are different values of cell and of ectoplasmic behavior. But there are general properties of the ectoplasm common to them all, properties that constitute the basis of vital activity: no living thing exists that does not display conduction, contraction and respiration, that does not take in food and water and eliminate effete matter. By virtue of conduction, cells in multicellular organisms are integrated.

This, then, is the general biology of that portion of the living protoplast located at the cell-boundary. Having reviewed it, we are now in position to discuss the ectoplasm as the primordium of nervous structure and function. This, in turn, will permit us to formulate our theory of the origin of man's ethical behavior.

Page 92

Chapter 5 – **THE PRIMORDIUM OF NERVOUS STRUCTURE AND FUNCTION (CONTINUED)** – pp. 92-100

It would seem from the fore-going that all vital phenomena reside in the cell-surface, a conception at variance to that more generally prevailing one, which emphasizes the dominance of the nucleus—where this is present—buried within the cytoplasm. Here, I lift a common-sense proposition, which no one would deny—that every living mass of protoplasm has a boundary—to the dignified level of a scientific hypothesis, and with it the no less undeniable and obvious fact that, by virtue of its location, the ectoplasm must carry out certain functions. It would thus appear that in this manner of a wholesale attributing of important functions to the cell-surface, I create a difficulty for the elaboration of my present proposition, namely, that the ectoplasm and its behavior are the primordium of nervous structure and function. Why should it not be, if the ectoplasm is everything? The entire multicellular organism is derived from a single egg—a sufficiently difficult problem; I would, apparently, increase the difficulty by assigning the major burden of development to one region of this minute mass of living substance.

To offset these possible objections is no onerous task.

Since here cells are dealt with, the occurrence of non-nucleated living things must not be over-emphasized; the present discussion is therefore limited to cells. Then to exaggerate the importance of the ectoplasm is to commit the same error of those who make the gene, a particle of the chromosome of the nucleus, the unit of life. It is just as true for the ectoplasm as for a nucleus, a chromosome or a single gene that it cannot

Page 93

stand alone. But my argument is not one for the independence and self-sufficiency of the ectoplasm, that it alone is the life-substance. Rather, the argument is this: that as the unit of life, the cell expresses certain tangible activities as visible displays of the cell-surface, the peripheral extension of the cytoplasm. In science, most of all in biology, the first step in the successive approximations to truth is taken when we appreciate phenomena, that which we can see. The ectoplasm is that region of the cell where visible expressions can best be grasped.

Now I know full well that there be learned masters to whom common sense is anathema, who hold that it is, or ought to be, under an interdict in science. These will certainly object to the obvious being brought within the sacrosanct walls; according to such learned ones, the secrets, mysteries, rites and tokens must be hidden away and protected from the intrusion of the uninitiated and his naive experiences. The statement that cells have boundaries that, because of location, carry out certain functions is far too banal, it might be said, to be called scientific. Why has the obvious been so long neglected, if not for this reason? Here again I would say that in biology a safe way to journey toward the truth is to start from the known, even the obvious.

It is very easy for most of us in view of its tremendous complexity to think of the nervous system in higher animals as something apart from other bodily structures; whereas of course its structural basis is the same, the cell. Its activities, even more complex, are also cellular. Neither its form nor its functions are of _de novo_ origin: in the evolution both of the animal kingdom and of the individual species of animals, the nervous system shows progressive development. Its origin must be traced back to the egg, for only multicellular organisms are endowed with nervous tissues.

The suggestion of seeming plausibility that I make difficult my case, that nervous function is a derivative of ectoplasmic behavior since I seem to make of the ectoplasm everything, is easily offset. Both ectoplasm and nervous function are related—they are unified in the simplest case through limitation, and in the most complex case, through integration; both directly relate organism and environment. The problem set, to untangle this behavior from others, contraction, etc., is not made more complicated by postulating a definite region in the egg whence arise these behaviors. If in positing functions to a given area we do not lose sight of the unity of the cell, this objection has no real weight.

So too the objection that might be made, namely, that I sub-divide the already sufficiently minute egg in assigning definite functions to its ectoplasm, can be likewise met. The development of functions, as that of structures, is a catenary process, in some measure autocatalytic. Hence their beginnings in the egg are to be correlated with that which increases progressively.

Let us consider any animal egg, from a sponge's to human included, asking ourselves the question, what, in the significant period for the laying down of the organ-systems, has increased? When the chick breaks through its shell, it does not contain more substance than was in the egg when laid. Although the mammalian young, when born, is of far greater weight and size than the egg at its beginning of development, fertilization, it does not increase

Page 95

during the period of the founding of the primary initia, the three germ-layers, of the organs-to-be. The development of many animal eggs is very much simpler than in these two examples. I have in Roscoff watched, for instance, the direct development of two species of sea-creatures which hatched out without metamorphosis in three to four days. Some other eggs, of low chordates, hatch as larval forms eight hours after fertilization. In these cases of complete development, increase of substance by addition from the outside is out of the question. To be sure, oxygen, indispensable for life, is taken in; but water in the meantime is lost. Thus, for any animal egg, it holds that there is no increase in total mass during the initial and most important stage of development, that which alone is common to all eggs.

What do increase during this period are total nuclear mass and ectoplasm. This period, of cleavage, during which the egg sunders itself, making itself progressively a many-chambered structure, is characterized by a transformation of nuclei and of ectoplasm directly from the hyaline cytoplasmic menstruum, with or without additions from the non-living bodies, oil and yolk, suspended in, or pendant to, the cytoplasm. The hyaline cytoplasm is of course progressively diminished in this process of the elaboration of nuclei and of ectoplasm.

Nuclei, as I have abundantly shown several times, cannot, in the light of the fact that they are elaborated out of the hyaline cytoplasm, be regarded as self-sufficient, unchanging governors of development. Less so their constituent chromosomes or the genes that compose these, especially since, according to the founders of the gene theory, in every cell throughout the life-history, the constituent genes remain the same. But if one rejects this view, one then must agree with Delage's logic according to which differences in nuclei are secondary, due to differences in their enclosing cytoplasm. Moreover, we are entirely in the dark as to the physiology of the nucleus; not one scintilla of evidence warrants such statements as that of Professor McClung that the chromosomes are the end-all and be-all of definitely named physiological processes; the statement made by W. B. Hardy that "the capacity for reproducing a particular molecular architecture in space which is called growth, is found in the nucleus", has yet to be proved. Thus we are led to conclude that the progressive increase during development of the ectoplasm, for which we have indisputable evidence of definite functions, has relation to the progressive differentiation whereby egg becomes embryo. More, this relation is definitely and clearly causal.

Development having been initiated by the spermatozoon or without it (natural or experimental parthenogenesis), the egg is lifted to a higher plane of livingness and, in a way that we have yet to fathom, becomes step by step something else, an embryo. Leaving the low ground of singleness, one cell, by successive sub-division of itself, becomes a myriad of cells—a mystery of the loss of an individuality to a new whole, for the multitude of cells is never a mere agglomerate but a unity. The development of the egg is thus a kind of material transcendentalism. For lack of a better expression, we may say that the lifting power inherent in the egg, the lever of its transformation, is a differential factor.

Here is no place to enter a plea for one or another of the several theories of development. True that "it is only when one has a theory that one can make progress"; it may even be true that

Page 97

"one can make great progress with a false theory". But the wealth of theories of development embarrasses us. Let us recount rather the facts.

First, mark you, the whole process of the egg's development is one of successive restrictions of capacity. As mentioned above, experiment having shown that each fragment obtained from an egg before fertilization is upon fertilization capable of developing into a complete embryo of size proportional to that of the fragment, the first restriction to occur in the normal intact egg, when its development is initiated, is that from poly-embryony to mono-embryony. As a mono-embryonic system, the egg by the repetitive process of sub-dividing itself is converted into a mass of adhering cells which at each succeeding cleavage have diminished power. It is as if in the fertilized egg were, say, a thousand twenty-four potency units, the whole number being necessary for a complete embryo, five hundred twelve being distributed to the two cells at first cleavage (ergo: segregation!) two hundred fifty-six at the next and so on until finally each cell has only one potency. This is, let it be understood, only a crude picture of what actually occurs; but it serves to make clear what I mean by saying that development entails a process of successive restriction, a progressive running down of potency.

By this picture I do not by any means intend to convey the impression that I subscribe to the old doctrine that the development of an animal is but an enlarging of the miniature of it in the egg, i.e., that the embryo is preformed or in the rough in the egg. No more am I proponent of the modern version of the preformation-theory, namely, that in the egg the parts of the embryo that are

Page 98

to be dismembered and scattered are, during development, properly ordered and segregated by some force comparable to the sound of Gabriel's trumpet which, according to certain religious believers, on the day of judgment summons the dead from their graves relating properly the scattered bones of each. Professor Ch. Perez has clearly and neatly exposed the fallacies of the preformation-doctrine, both the ancient and the modern. Rather, development is an unfolding of chemical processes in the egg reacting to the environment, as Delage suggested—in Professor Wintrebert's terms, a physiological epigenesis.

This progressive loss in potency appears to be correlated with the progressive diminution in size of the cells. It is, however, not to be attributed to mere subdivision of the egg. To subdivide a fertilized egg does not bring about its development. On the other hand, an egg in which cleavage is suppressed can develop into a moving embryo, albeit highly pathological, with an ectoplasm portraying exaggerated and abnormal activity. What is significant for normal development is the distribution of ectoplasm to the constituent sub-divided chambers of the egg; each receives its peculiar partition. As with repeated sub-divisions of any mass, the total surface-area increases, so with the egg, which steadily converts endoplasm into ectoplasm. By this and by the steady growth of nuclear stuff, the internal cytoplasm tends to exhaustion but never attained inasmuch as the utilization of food restores the depleted protoplasmic forces.

The first phase of development, the laying down of ectoderm and endoderm in all animal eggs and of mesoderm in those that possess it, is characterized by the distribution of the ectoplasm

Page 99

on the egg at the moment of the initiation of its development. Although the origin of endoderm and mesoderm differs somewhat in different species of eggs, the cells of these two layers arise from the less active regions of the egg: from the lower hemisphere of the egg, incorporating the ectoplasm of this region; from cells beneath the egg's ectoplasm; or from cells beneath the ectoderm, as in the case of the mesoderm of the chick's egg. The ectoderm, on the other hand, always arises from the upper and more active hemisphere of the egg; its cells are always the most superficially located and always include the largest quantity of the original egg-ectoplasm.

As stated earlier, the nervous system is invariably of ectodermal origin, having developed from the superficially located cells of the egg. Hence, its primordium is the egg's ectoplasm. During development, just as ectoplasmic structure is partitioned off to the constituent cells, so ectoplasmic functions are distributed in such wise that in the cells, retaining all general properties, one or another of these is emphasized; as contraction is most highly displayed by muscle-cells, which are mesodermal origin, the power of conduction is enhanced in nerve-cells. These are incontrovertible facts and not hypotheses or theories.

Man's nervous system, as that of any animal endowed with such, is the evolutionary product of the most active peripheralized cytoplasm of the egg. Its functions are exquisite refinements of the behavior of protozoon's and egg's surface concerned in the binding of the regions of the organism, making it a unit, and in the bringing of the whole organism _en_ _rapport_ with its environment

Page 100

How precisely out of the tangle of the general properties of the ectoplasm that to become nervous function is separated, is yet to be known. To say that in the course of evolution division of labor emerges is to describe and not to explain the fact of this separation— one never complete since all living cells retain and must retain the general properties basic to the life-state. So, in potential, the tangle remains. Moreover, as we have seen, the action of the nervous system in integrating regions of the organism is but a refinement of the more primitive cell to cell integration; other ectodermal cells than nerve situated on the surface of an organism, with or without nervous system, relate organism to environment. To understand the functions of man's nervous system, we need ever to bear in mind that they have descended from the simple manifestations of organisms in the single-cell state; that they bear yet the mark of the more primitive relationship exhibited by all living beings.

Page 101

Chapter 6 - **MAN AND THE OUTSIDE WORLD** – pp. 101- 123

The striking adaptations of many animals and plants to their surroundings are in most instances so marvelous that they justifiably provoke expressions of wonder, the more so as we fail to find a satisfactory explanation of these exquisitely refined adjustments upon which the continuity of the species depend. Recall the hermit crab housed in a shell, previously inhabited by a snail. How does this hermit crab know to protect its abdomen that in the course of evolution has lost the hard protective covering characteristic of crabs, by inhabiting an empty shell? And what about the sea anemone, a flower-like animal that affixes itself to this shell, becoming the partner in the life of the crab? Numerous cases of adaptations among animals and plants are puzzling as this.

Perhaps hitherto in our attempts to explain these "fittings" of organisms and environment we have erred in placing too great emphasis on the role of natural selection which for many biologists is the _deus_ _ex_ _machina_ for all problems growing out of the concept of evolution. Natural selection, or Darwinism, being not proved, we ought re-orient our thinking, and approach the problem of these curious modes of adaptation from another point of view. Certainly, we should not in emphasizing these peculiar, in some cases even bizarre, expressions of adaptations—the very term adaptation, implies a preconceived notion, perhaps even a misconception—fail to give due weight to the common factors in the relation of organism to environment.

Page 102

Upon life, the thread common to all living things, is spun the variegated patterns that together make up the world of animal and plant species. As we have seen, living organisms display certain activities common to them all; nevertheless, they retain and transmit each its own pattern, each true to its species. For every animal, life as abstract principle, Bios, the common property of all living things, is basic to the life of the individual species as such.

The inorganic environment also is a variegated pattern: of oceans, seas, lakes and rivers; of mountains, valleys and plains; of continents and islands. Extending from pole to pole, it is of many zones of varying climates each with seasonal changes. And over all, the sun is hung, giving an additional nuance. Even the clouds contribute to the pattern, making each region different from all others. Under the sky of Paris every man is a poet unawares. Nevertheless, beneath this endless variety of the nonliving scene are factors common to it all: water, gases and minerals. Any region of the inorganic environment may be regarded as but a combination of these constituents under the light and warmth of the sun.

Life does not exist apart from these factors common to the inorganic environment. Water, quantitatively the most important compound in this environment, holds the same place in the constitution of the living animal or plant. Oxygen is indispensable for life-processes. The nitrogen cycle familiar to all indicates the utilization of this the most abundant gas of the atmosphere. Plants convert carbon dioxide plus water in the presence of sunlight into foodstuff. Upon this process, photosynthesis, the

Page 103

animal kingdom depends. Thus, we are creatures of the sun. Protoplasm, plant or animal, is a water-laden substance never free of mineral salts. The relation between organism and the inorganic environment is therefore that of two regions in nature having factors in common.

From this point of view, it would seem gratuitous to speak of the adaptation of organism to the inorganic environment as if the relation were due to qualities superposed upon the organism, an after-thought, as it were, on nature's part. The often spoken of "fitness" of organism to environment, I think therefore a maladroit statement. So too the idea of the "fitness" of the environment. Neither organism nor environment was separately measured and cut to fit the one or the other. It has always seemed to me that in these conceptions of fitness inhere a strong undercurrent on the part of their evolutionist authors of thinking in terms of special creation—of the best of all possible worlds and of organisms snugly fitted to it. Certainly the latter conception, of the fitness of the environment, is an argument for design in nature, of the creation of the universe for the purpose of the living world with man the highest achievement of the creator. Of course, nothing can be said against this philosophy if it be proffered as avowedly that of special creation. By the same token, it lacks force, and logic, if it comes from an upholder of the theory of evolution.

The concept of evolution ought not be limited to the problem of the origin of now existing species; it should include also the question, how life first arose. To be sure, we can ignore this question, and fall back on the idea of life coming from another planet,

Page 104

or postulate an initially created being with subsequent evolution. However, the postulate of the genesis of life out of the nonliving environment existing at the time of the origin of life appeals to me as a logical necessity for the thoroughgoing evolutionists. Life having evolved from the environment, part of which composes the organism, maintained strictly the relation to it as source. Even today, every living thing in a sense "comes from" the environment by the water, gases and electrolytes that it takes in. So too with respect to its food. A living thing today persists and is capable of maintaining itself because it incorporates part of the environment, just as its original ancestor, the primordial form of life, came to be by the partitioning off of organism from environment. Organism plus environment—this is neither a fitting nor an adapting, less a kind of symbiosis or parasitism; it is a unity. Life, even for the longest-lived organism, is that brief moment in eternal time when organism and environment appear as a dual system of reversible change, the reaction tending to run in one direction. It is the moment of a single trembling flash before equilibrium and the subsequent run of the reaction to the side of the environment. Individual life is but the hesitant step between its beginning and its end. So brief is this compared with the long history of living things on this planet that we can imagine that, could we see the whole procession of living things since their origin, we could not distinguish between organism and environment.

Along the thread of then-common factors present before life emerged we trace the evolution of the inorganic environment to its source: the earth and the fullness of life thereof are the

Page 105

sun's. And they have never escaped this origin. Under the dominion of the sun these factors became sequestered into what we name animals and plants. Might it not be that the intermediate steps in the welter of processes that together comprise a living thing escape us because we have come to think too much of organisms as being static because of their maintenance of outward form whereas these processes run with the speed of light?

The relation between organism and environment was thus not, as the terms, fitness and adaptation, imply—even denote—established subsequent to the origin of life. It is for members of the living world the oldest relation there is—that which called them into being, that which makes possible their continued existence. The original partitioning off of part of the environment from the remainder, by which life came to be, must have been a crisis in the drama of this planet. Somehow emerged a differential without which the integrity of the newly arisen prototype of the animal and plant kingdoms could not have persisted. This retained a large communion with the not living portion of the environment while establishing the bounding limit of the living thing and conditioning its activity.

The fore-going statement is not intended to overlook or even to minimize the aspect of inanimate nature hostile to organisms. Factors that compose the inorganic world are in certain combinations lethal to living things. Poisonous gases exist; carbon dioxide utile for green plants is toxic to animals and even oxygen may be deadly, other single gases being more dangerous. Some minerals kill, and water so indispensable to life may end it. Outside a certain range of temperature life is no longer supported. And so on. Notwithstanding all this evidence of the harshness of inanimate nature, life has persisted since its first coming into being; without being en rapport with the environment, life could not have sustained the shocks that have marked the history of this planet. We are justified then in emphasizing the relation of organism and environment as one of interdependence. Without regarding inorganic nature as a kind and gentle mother showering benefits upon her offspring, living things, we need not go to the other extreme, the position maintained by many evolutionists, namely, that perpetual war is waged between organisms and that implacable foe, the environment.

What strikes me as the more significant is the interdependence, and not the antagonism, between organism and environment, meaning now not only the inorganic but also the organic. In the moment that one speaks of war between organism and environment, it becomes imperative to include the term, environment, all nature external to the organism inasmuch as in employing the expressions, struggle for existence and survival of the fittest, etc., one usually refers more to the living than to the non-living world. Indeed these phrases, originally used figuratively, came, as debased currency, to lose their stamp, to mean war to the death not only among species but also within species. The tag, "Nature red in tooth and claw" [Tennyson], became both the shibboleth and the apotheosis of Darwinism. The theory of natural selection was now generally accepted as the plot of a grandly horrific spectacle of immolation.

With the emergence of life, the primordial environment changed. Thenceforth it was incorporated of inorganic and organic factors.

Page 107

These latter were derived from the presence of the living thing itself, its end-products during the lifetime and those substances yielded up in its death and disintegration. This alteration in the environment increased with each succeeding step in the evolution of the natural world. And these steps were not reversed. With the first respiration of the primordial living thing, its energy-changes, its elimination of wastes and finally its dissolution in death, the external world changed never to return to its pristine state antedating the advent of life. It moved ever in one direction. Its retrogressive movement is as difficult for us to conceive as a return of this earth to the sun.

Thus, slowly and steadily was built up the organic environment of organism to its present state. To this environment animals are related dependently and interdependently. The dependence may be direct or indirect, as, for example, that upon plants, which maintain stricter relation to the non-living organic world than animals. Many animals utilize directly plants for food; but in the final analysis the animal kingdom, including its flesh-eating species, is helplessly bound to green plants by virtue of their capacity to synthesize their food and their protoplasm from chemical elements and simple compounds. The relation of animals to plants may also be spoken of as favorable and unfavorable, there being plants that are beneficial and others injurious to animals. The relation of animals to animals may be that of species to species or that among members of one species. Because these two relationships are different they ought be discussed separately.

One relation of animals to other species of animals is dictated

Page 108

by food requirements. Whilst it is true that herbivorous animals nourish themselves on plants, the carnivorous feed upon the herbivorous. In the words of Claude Bernard, founder of modern animal physiology and of experimental medicine, speaking of these two kinds of feeders:

> "These results which assure cosmic equilibrium are the consequences of the general law of the struggle for existence according to which nature can engender life only by death, creation by destruction. … Living beings can exist only with the materials of other beings dead before them or destroyed by them. Such is the law." [Claude Bernard 1927 (1865 in French)]

Were all carnivorous animals suddenly converted into herbivorous, the sustenance of the animal kingdom would still exact destruction of living things.

Here is not only a hostile relationship but an enmity necessary for the existence of those that destroy. This is certainly a struggle for existence—on the one side by those that would escape and on the other by those that would kill. Nor is this struggle limited to animals: the lilies of the field must toil and spin or lose the thread of life. Moreover, in every living cell "the grand cosmic phenomena" are manifested, here also in nutrition is a struggle against disintegration; the moment that the processes of wear and tear overcome the reparative, life ceases to be. Struggle, be it aggression or merely physiological work, to sustain life is written large in the history of the living world. And all our sentimentalizing over nature cannot blot out a line of it.

Nevertheless, one may voice a protest against the exaggerated use by many evolutionists of the term, struggle for existence, employed by Darwin in a metaphorical sense; one can indeed even in

Page 109

this original usage detect weaknesses in it as a concept of a causative factor of evolution.

Throughout all nature prevail action and interaction tending to produce equilibrium, from which one comes to speak of the cooperation and harmony of nature. Here we ought avoid confusing the actions and interactions in inanimate nature with those in animate nature. The former are due to forces analyzable as purely physico-chemical; in the latter, in addition inhere other forces which as yet escape physico-chemical analysis. Inanimate matter being devoid of will, consciousness and purpose, and living substance possessing these as such or displaying a behavior that is their primordium, the actions and interactions in inanimate and animate bodies are again different. Non-living things very closely obey those laws, even a very slight displacement being sufficient for its complete destruction following which energy-and matter-changes are now of a different order. Quite apart from its greater complexity and heterogeneity, living protoplasm portrays a peripheralization both in structure and in function that sets it off sharply from a non-living system, a peripheralization which is the prevailing characteristic of the life-state and conditions its actions and interactions with the surroundings from which actually never in experience it can it be separated. Into the composition of protoplasm enter materials of non-life—any other proposition is scientifically inadmissible; but protoplasmic organization does not rest solely upon the combinations of these as elements and compounds; it rest also upon the order of these combinations in space and in time, an order peculiar to living substance. And this we can never understand by the mere transference

Page 110

of physical, and less of mechanical, concepts to life. All such attempts have up to now failed. Hence, in several directions, the ground-postulate for the physical sciences, that which concerns the relations between inanimate bodies, has not yet been extended, and in my judgment never will be extended, to embrace the actions and interactions of living things.

Now equilibrium (sequus, equal and libra, balance) here means the state of repose of a body due to the offsetting of forces or destruction. Let us extend the literal meaning of the term, dismissing its physical connotation as applied to inorganic bodies, to include cooperation and harmony in the living world, as, for example, the nitrogen cycle interposed between animal and plant life. But in neither meaning, of a counterbalancing of destructive force or of cyclical changes, can equilibrium be reconciled with the forward upward process of evolution. A struggle against destruction would result merely in a check, a maintenance of a certain level, for if it outweigh destruction, then there would be no more equilibrium and a cyclical change. A revolving of this kind, through the utilization of the available quantity needed, does not imply an evolving. Either in the more popular interpretation of the Darwinian concept of the struggle for existence as war to the death or in Darwin's own metaphorical usage, even with the auxiliary postulate of variations seized upon by natural selection, that which would be the fittest to survive could survive only; there would be no evolution.

An extended discussion of the weaknesses of the theory of natural selection, so clearly exposed by many writers, would be out of place here. That there is by every living thing a struggle—

Page 111

be it fight, work, or relatively passive intake—to maintain life and that species of animals and plants vary, none will deny. Nonetheless we wonder why the "struggle for existence" was made the mainspring of Darwin's theory.

The ideas put forward by T.R. Malthus in his book, <u>Essay</u> <u>on</u> <u>Population</u>, which inspired both Wallace and Darwin, according to which human population would increase to such numbers that it would lack adequate food-supply have, thanks to modern methods of production, been so far disproved that we have lived to see the deliberate and wanton destruction of animals and plants used for human needs in accordance with the doctrine of scarcity. In America at least, the New Deal has thus, with the Wallacian, jeered the Malthusian doctrine. However, it does not follow from this that Malthus' doctrine, where it concerns checks to over-population, applied by the co-founders of the theory of natural selection to animals and plants, is false; rather, it applies more aptly than to the human populations. Even so it is false. Transferred again to human society it has been a weapon for evil, especially as used by the cults of the war-making guilds.

On the other hand, because one rejects this re-transferred doctrine, one need not accept the Marxian view, in so far as this latter interprets the class struggle as the form in which Darwinian struggle for existence is extended into human society. It does not follow that because the Darwinian creed is wrong, the Marxian view has to be right. The transfer of power to the proletariat does not seem to mitigate the class struggle, if we can accept the evidence drawn from the large-scale practice to which this doctrine has been put, except by exterminating members of other classes and those of its

Page 112

own. Of course, here is no class struggle—only class-dominance now in a larger measure than ever before. Add the fact that this communism, as practiced, discards the one form of communism that is ideal and as yet unrealizable but not unattainable, Christianity.

The struggle against, or the hostility to, other organisms by no means represents all the relations of animals to other animals and plants. Some others may be mutually beneficial. There are, in addition, such modes of living together as commensualism and symbiosis. Isolated instances of care and of aid, as that cited by Eckermann—the story of a fledgling which, having fallen from its nest after its mother was killed, was picked up by the mother of another species—have been stressed by Kropotkin. But these, not being widespread facts, do not explain "the divine in nature" [Kropotkin 1924: 265 (Kropotkin quoting Johann Peter Eckermann quoting Goethe)]. Less are they to be considered a factor in evolution.

When we come to appraise the relations of members of the same species of animals to each other, we find it necessary to take into consideration common factors that modify interspecific relationships so far spoken of. Commonality of descent, of structural and functional organization, and of dependence upon the environment enter into play. Members of the same species considered en bloc, apart from incidental variations and individual differences, constitute a group which in the state of nature is set off from all others. For no matter what concept of species we support— of species as fluid and transformable or as fixed and unalterable—we never encounter outside of laboratory (i.e., artificial breeding) a break-down of the barriers that separate one species from another. Noteworthy experimental results on crossing species of the same and of even different genera have been obtained among plants, the offspring of the crosses being fertile. These results are in consequence of previous changes induced by cultivation or domestication, plants being more labile than animals. Thus, many plants, as the banana for example, lose their capacity for sexual reproduction as long as they are under cultivation. It is well established that cross-fertilization between even very closely related species of animals, if it succeeds at all, does so only after strong experimental means are employed to break down the specificity of animal eggs, a characteristic that greatly restricts, or limits, their reactions at fertilization to species-specific spermatozoa. Despite experimental results on cross-fertilization, species in nature remain true no matter how long they are closely associated with others. One may state this as a rule. There is today no transformation, experimental or otherwise, of species on such scale that warrants the statement that normally, in nature, reproduction giving rise to fertile offspring escapes the limits of the species-group. Thus, reproduction, without which there is no continuity of any life-form, lacking which there would have been no evolution of living things, no scattering of organisms in either time or space, maintains also the limits of species, makes of each a unitary group.

Obviously, an organism whose mode of reproduction is asexual, i.e., by the sundering of itself, is confined to a definite locality during the act of reproduction. For the purpose of sexual reproduction, the sexes of animal species congregate even though the male and female sexual products are emitted, without sexual congress, to the external medium, where they meet by chance.

Page 114

(This chance is less the farther away the sexes are from each other during breeding season because the sexual products are short-lived). Thus, reproduction tends to relate animal species seasonally to a given medium. The natural history of animals bounds with dramatic incidents of this dependence, as shown, for example, in the migration of the shad, the salmon, the eel and other fish.

The food required is also a factor in the dispersal of a species. Having the same needs, in some cases very definitely limited to a certain kind of food, a species tends to distribute itself according to the availability of the means of subsistence required. The demand for nourishment dictated by specific structural and functional organization especially by that of its digestive apparatus, together with the no less imperious need for inorganic constituents, water, oxygen and mineral salts, which common though they be to all living things are quantitatively satisfied differently in each species, is a factor in determining the relation of members of a species to the environment.

Thus, breeding and feeding both contribute to the localizing of a species. But of course, as we well know, the locations may not be identical, are indeed often quite dissimilar, as shown by the migration of fish mentioned above. Where migration antecedent to feeding occurs, it is periodic and may result in the aggregation of the species in a well defined and even a narrowly limited space. Movements in quest for food are in general more diffuse both as to time and to space. Thus, reproduction and nutrition act differently in relating members of a species to the environment.

These two physiological needs also relate members of species

Page 115

to each other, although the respective relationships differ.

For certain sessile animals, as sponges, some coelenterates, Bryozoa and some tunicates (low chordates), by virtue of their organization into colonies, living together is an inescapable mode during the adult (and major) part of the life-history. In such colonial forms, usually the sexes are united; hence in these forms, unless for reproduction, the eggs and spermatozoa from different individuals are mandatory—that is, if the sexual products elaborated by one individual cannot react with each other to induce development—the aggregation of individuals to form a colony cannot be attributed to the necessity of approximating the sexes. Similarly, if freely locomotor hermaphroditic animals tend to live in groups, this cannot be ascribed to the physiological need of reproduction; the failure of the induction of development in eggs by spermatozoa of the same individual of this or any other type of hermaphroditism in animals being far less prevalent than is often assumed.

This note on sessile and hermaphroditic animals is, in a way, beside the point of the argument: The factor underlying a society, a grouping, of sessile forms is environmental and has no attribute of the "herd instinct", such as that of the basis of a community of actively locomotor animals, as birds or mammals, that live together in flocks or herds. For, although as independent, freely swimming embryonic forms, sessile organisms may settle down in groups, we do not extend the meaning of the term, instinctive, to include the reactions of embryos which may come to a common place of settling down before they have developed a nervous system—some, as the sponges, never have nervous system. If the assembling of

Page 116

the embryos is held to be tropistic, then this is in relation to the environment and is not of organism to organism. An hermaphroditic animal is, from the point of view of reproduction, already social, segregated by having sequestered in itself both sexes. And yet it is perhaps well to make this note, especially with respect to hermaphroditism, because there is the rather popular notion that the incorporation of both sexes in one individual is abnormal for animals. When we bear in mind that the first group (phylum) of multicellular animals, the sponges, are hermaphroditic and that every other phylum includes some normally hermaphroditic forms, we ought no longer regard the separation of the sexes as alone normal. Hermaphroditism is abnormal or pathological only in animals in which the sexes are normally separate. Rather than attempting to explain away hermaphroditism, so widespread and normal an occurrence in the animal kingdom, we need to explain how and why the separation of the sexes arose. We encounter very much the same question in the problem of fertilization: Since normal parthenogenesis (development of the egg without contact with the spermatozoon) takes place, albeit not in widespread occurrence, and since eggs for which normally the stimulation by the spermatozoon is necessary, are capable of the experimental induction of development, how and why did fertilization come to be? Less wonderful than the notion of the "immaculate conception" of a fish or a frog—there is no valid reason against its being possible in mammals including man although biologically the difficulties are great—is the fact that fertilization has to occur.

The coming together of free-living animals of a species in which the sexes are separate is made mandatory by reproduction.

Page 117

The most solitude-loving animals require propinquity for reproduction, the surest sign of social equality. However sexual attraction is analyzed, as instinct, as due to sense of smell, etc., the sexual relation, except for normally parthenogenetic forms, is in nature an imperative that is antecedent to the propagation of the species. Sexual reproduction in this category of animals, by necessitating that the sex-to-sex relation be localized, falls in with that of sessile and of hermaphroditic animals. It also falls in with asexual modes of reproduction.

Nutrition does not, to the extent as does reproduction, demand communion of the members of a species. True, there are innumerable cases of the feeding of the young by the parents and some of the feeding to the young of a parent—as among insects, the male is used for food—but these, as well as instances of other forms of parental care, are offset by the more abundant examples in the animal kingdom of neglect of the offspring. We cannot account for the segregation of animals in search for food in terms of an attraction between members of the group. The food is attraction to the individuals, acting now in their responses as a collection of individuals. What makes the collection a unit is not the quest for food, however exacting the need for a special diet may be, but rather those characteristics which makes it and preserves it as a species distinct from all other animals. Reproduction, as has been said, is the sine qua non not only of the perpetuation of the species but also of the continuity of both animal and plant life and is a factor in evolution. Animals eat to live and live to reproduce. Nevertheless, the need for food is the more powerful drive in the individual life. This cannot be denied. There exist normally sterile

Page 118

animals incapable of reproduction, as, for example, the worker-bees. Experimentally, animals can be rendered sterile: by castration life is not ended. But an organism deprived of nutrition dies. However, the quest for food is individual—as individual as the need of it is.

Although the fore-going exposition by no means exhausts the question of the relations of members of a species to each other, it suffices to show that the relations are closer than those between the species and other animals. They are, moreover, fundamental in that they hold for every species of animals, a point that especially one who theorizes concerning the "herd instinct" or the practice of mutual aid and cooperation—types of behavior by no means common to all animals—should appreciate. I return to this point later. To what has been said to indicate that these intraspecific relations are, in addition, different from the interspecific relations, another point ought be mentioned. I refer again to the notion of the struggle for existence.

The struggle for existence in the popular meaning of war to the death cannot apply to members of a species in the same measure as to members of different species. Even between different species of animals or between animals and plants, exterminative warfare is unthinkable: the extinction by a predacious organism of a form necessary for its existence would in the end result in the demise of the predatory form. Intraspecific war to the death, even if taken alone, would be disastrous for the continuity of the species; added to warfare from without, it would mean rapid annihilation. To grant the existence of some warfare between species does not entail the assumption that intraspecific warfare must also take place.

Page 119

True, in time of stress, a struggle to death may be waged among members of a species: for example, normally non-cannibalistic organisms may become cannibalistic when their proper food is lacking; some animals are normally cannibalistic; instances abound of fight to the death between members of a species. It would be idle to deny these facts. What I insist upon is that the doctrine of extermination is incompatible with the mere survival of the species. This popular meaning ascribed to the struggle for existence ought not to be extended to include species-suicide. There is no such extension; indeed, the tangible evidence indicates that species have survived. (though not all!)

Consider, then, the struggle for existence as a war of less magnitude among members of a species, together with variation within the species, to the end that the fittest selected by nature survive. Intraspecific battling, although now not leading to (extinction) extermination, would attenuate the species whose variations, not being beyond definite limits, would scarcely be such with respect to fighting-prowess that some members would far surpass the others; even if there were those super-members, the hazards of war might not save them after the "blood-baths". The postulate of intraspecific warfare as a cause of evolution is to me unattractive because illogical. The world of living things as now constituted could not have come to be as the result of a descent ordered by war: the war-complex would need to be handed on from generation to generation with none or few to receive the heritage.

Incontrovertible are the facts of the struggle for existence in the metaphorical sense as used by Darwin and of the variation

124

within each species. Add to these the fact that evolution has taken place. Granted, all this does not give answer to the question: How did the species now in existence come to so closely maintain themselves, each within limits? Granted that this struggle for existence, having weeded out the unfit among the variations, would lead to the survival of the fittest. Even so, this natural selection would mean survival only, and would not account for the fact of the progressive ascent in the living world.

The idea of sociality, first advanced by Darwin and developed later especially by Kropotkin, as a factor in evolution, was vigorously emphasized by the latter to offset this popular interpretation of the struggle for existence as nature's law of war for all against all. Kropotkin saw in his concept of mutual aid and cooperation "an inspiring truth":

> "There is still less foundation for another continually repeated reproach to empirical thought— namely, that the study of Nature can only lead us to knowledge of some cold and mathematical truth, but that such truths have little effect upon our actions. The study of Nature, we are told, can at best inspire us with the love of truth; but the inspiration for higher emotions, such as that of "infinite goodness", can be given only by religion. It can be easily shown that this contention is not based on any facts and is, therefore, utterly fallacious. To begin with, love of truth is already one half—the better half—of all ethical teaching. Intelligent religious people understand this very well. As to the conception of "good" and striving for it, the "truth" which we have just mentioned, .i.e., the recognition of mutual aid as the fundamental feature of life, is certainly an inspiring truth, which surely will some day find its expression in the poetry of Nature, for it parts to our conception of Nature an additional humanitarian touch."

[Peter Kropotkin, (1924). *Ethics: Origin and Development*, p. 25]

The idea here expressed, of imparting to the conception nature a humanitarian touch, is, I think, unfortunate for a scientific treatise by a scientist; it is doubly so in the connection used. But it indicates Kropotkin's conviction of the truth and beauty of the conception of mutual aid—a conviction so strong that it sometimes swept him beyond the limits set by the evidence. Nevertheless, examples of mutual aid and cooperation

Page 121

within species sufficiently abound to warrant the statement that it is far more prevalent than intraspecific warfare. As I have said elsewhere, the Kropotkin theory may be, therefore, a better explanation of the cause of evolution than this popular notion of a war to the death. Kropotkin himself subscribed to the Darwinian theory of natural selection. If, however, we reject the theory of natural selection, the evidence of mutual aid and cooperation, although by no means found throughout the animal kingdom, still remains. What to me is more important than the examples of such practice is the basis whence springs the practice, especially if this basis be in some manifestation which is common to all members of the animal kingdom and which is consonant with the arrangement of species in an ascending linear series; for the practice of mutual aid and cooperation in the Darwinian or Kropotkin sense is certainly not common to all species of animals. From this point of view, whence I envisage it, cooperation has to be reckoned with as a contributing factor to the cause of evolution, whatever may be the theory of the cause.

The organism is product of its environment—arose from it, depends upon it, returns to it. From gases, water and salts, warmed and lighted by the sun, came the first living thing. A breath, water-laden, and the salt of the earth sustained it. As dust it returned to dust. In the unceasing upward progress of the living world, life, eternal since it began, has but inscribed a wider circle, unending repetition of this first cycle; each individual life-form is but an unvarying repetition of this first cycle: Back and forth between organism and environment course streams, and where for a moment there is an eddy, a seeming pause,

Page 122

life sits. The adaptations of life to environment are minor fluctuations in the velocity and pressure of the flood; they are softened accents now of the mighty cataracts of past ages which shifted the current but never altered the course.

Once life emerged, the environment changed and with each stage in evolution changed farther, the fundamental relations between organism and environment persisting with the added ones of animals to plants, animals to animals of different and of the same species. By virtue of the processes of reproduction and of nutrition peculiar relationships exist between members of the same species. These processes tend to bring about the aggregation species, the aggregations, however, having different value: the quest for food is selfish, being for maintenance of the individual; the impulse for reproduction is the one unselfish urge portrayed by animals for the perpetuation of the species.

Man as animal has like relationships to his outside world. He needs oxygen, water and electrolytes, depends upon the light and heat of the sun. He has enemies among both animals and plants; he derives his food from both. Tracing his lineage to the first form of life, he has inherited from his ancestors. But he is also the apex of the living world; as such his relationships are more acute. As most skilled hand-worker, he is above the brute-world; his relationships are finer. As thinker by virtue of possessing the acme in development of primitive ectoplasm, he penetrates his outside world in time and in space; his spirit gives meaning to his own and to all relations in the universe.

Page 123

Man is the crowning achievement of the living world. The lowly first slime of life, freed from the primordial ooze, held the spark of his being. A creation of this kindled spark, he is yet lord of it. Out of the immense, long shadows of the past he has emerged, glowing with this inner light, the effulgent glory of the ages.

Within the purview of all that has been heretofore set forth, I envisage man's ethical behavior as an expression of his organization as animal and as man, an inherited behavior within the larger framework of the relation of organism to surroundings refined through evolution of man to fellow-man.

Page 124

Chapter 7 - **THE HUMAN MIND** – pp. 124-149

Although composed of a myriad of cells, man's body is a unit in structure and in function. A human being lacking one or all sense-organs, one or all four limbs, one kidney, the entire spleen, all the organs of the sexual apparatus, or some portions of other organs—or all these parts named—can maintain his being, albeit with diminished efficiency. For unlike a man-made machine, as an automobile, man, as all organisms, can make adjustment after serious loss and is capable of self-regulation, whereas the automobile is not. With the reservation necessary in view of the possibility of existence after even severe mutilation, the human organism is a unit.

Among those myriads of cells are the sex-cells, the ova of the female and the spermatozoa of the male—elaborated, respectively, by the ovary and the testis, organs without which their possessors can live—which are virtually independent of the body. All other cells also have independence, though this is masked, as shown by their capacity to maintain existence when cultured in vitro (outside the body). Independence is a characteristic of human cells as of all cells, animal and plant. But whereas the sex cells, destined to make up a new individual existing free from the parents, emphasize this fundamental cellular characteristic, the other, the body or somatic, cells, except in cases of cell-anarchy as in cancer, remain more strictly en rapport with each other.

The myriads of cells, including the sex cells, comprise a

Page 125

community transformed into a unity through the coordination of inherently independent cellular activities. Structurally, the body would be a mere aggregate of cells but for the interdigitation of ectoplasmic processes that establishes a close knitting of cell with cell. Since this interlocking of cells underlies all kinds of integration by structure, including nervous, and is thereby the means of establishing coordinated and harmonious activities of the cells, the ectoplasmic processes subserve not only the structural but also the functional unity of the human body.

Thus regarded, this unity appears as derivative, sustained by the abnegation of independence on the part of each cell. It seems to be conditional upon the community of cells at its basis. But this community itself, in the human body as in every other multicellular organism, derives from a single cell, the egg, an independent unit which by a process of successive autotomy converts itself into a community of cells. The prototype of the animal egg is the protozoan cell, a more self-sufficient and more enduring independent unit. Both in its individual history and in its evolution, therefore, the unity of the human organism is discovered as antecedent to its state as community of cells. From this point of view, then, unity becomes paramount and the community derivative.

Regarded in the light of its history, the multicellular organism, man included, is the creation of a single cell, the egg. But if this history is overlooked, one may entertain quite another view. For example, Professor H. W. Shimer, the eminent geologist has recently said:

Page 126

"Late single cells came together to form many-celled larger units with their enlarged possibilities. In each of these the component cells divided the division of labor of preserving the larger organism, and hence the existence of each individual cell came to depend on the existence of the other cells. And it is possible that in these larger many-celled organisms may have arisen the first diffused beginnings of the impulses which are sometime classified together as the herd instinct."
[No source cited. Also, see *Evolution and Man* (1929) and *Introduction to the Study of Fossils* (1933) by Hervey Woodburn Shimer.]

Each of the sentences quoted is susceptible to adverse criticism. For our purpose, however, the following comment will suffice:

The established date for the development of multicellular organisms proscribes the statement, expressed as simple fact, that multicellular organisms originated from an assembly of single cells. In the first place, animals may, in nature or as the effect of experimental treatment, develop parthenogenetically—i.e., from eggs alone, without the intervention of the spermatozoon. Further, there exist cases where, although the union of egg and spermatozoon is obligatory, the spermatozoon is inert. More, in all other types of fertilization, the most widely occurring being one in which the spermatozoa take the most active part known, development ensues when and only when the spermatozoon loses its identity to the egg. Thus, development consequent to and without the coming together of egg and spermatozoon is always a function of the egg. Waive all these considerations, a most serious objection to Professor Shimer's statement remains, for an animal egg always develops by an initial sub-division of itself, making successively smaller chambers of the stuff intrinsic to it. Only later through the processes of nutrition does it grow in size, becoming a larger unit than was the egg at the initiation of development. I limit this comment to multicellular animals because I imagine that the writer, by speaking of herd-instinct, diffused beginnings or otherwise, excludes plants. Likewise

Page 127

Protozoa, being outside the sub-kingdom, multicellular animals, pass unnoticed.

That the conclusion, the possible rise of the herd-instinct, is a complete non sequitur, may be overlooked. More germane to our argument then this logical fallacy is the simple declaration as fact that which is contrary to fact. For no structure or function arises from the complex, qua multicellular organism, conceived as the resultant of the association of even two single cells since no such complex originates thus. As the associated cells, so too the means of association and the association itself in the developing complex both result from the self-fabricating power of the egg. Far from being an additive compound of single units, the embryo and later the adult is the creation by a single unit of a time-ordered translocation of space of a progressively sub-divided substance, wholly endogenous to it, so distributed as intercommunicating chambers that their repeatedly altered dispositions—both with respect to each other and to the outside world, whence came the stimulus, spermatozoon or experimental agent, that kindled the spark of the creative process— evoke, successively new reactions and responses which culminate in those peculiar to and denominative of a given organism.

The unity of the human, as that of other organisms, is not an acquisition; it has not arisen through suppression of previously independent members of the cell-community; nor yet is it a vestigial or atavistic character. Rather, I conceive it as a persisting reality, fundamental to protoplasmic organization. Nonetheless, the concept of the body as a community of cells is inescapable; it too is a reality although of less widespread occurrence, being

Page 128

limited to multicellular organisms. There is here no contradiction, provided we bear in mind the genesis of the cell-community. That is to say, communion among cells is no less real than the aboriginal unity because the means of communion among the cells are themselves products of a single cell, the egg. What I am insisting upon is the important role of these means, the ectoplasm, in maintaining the enduring reality of protoplasm, once primitive ectoplasms had partitioned off from the environment that which became the enduring reality. Ectoplasmic structure and function are essential to protoplasmic organization. From egg to adult, the process of development is as much the laying down of intercellular connections as it is the setting up of cells.

As community of cells, the multicellular organism expresses the state of being alive only through the manifestations of the individual cells. As a community of cells, the cellular interlacings retain the role of impressing the persistent reality. As community of cells, man, the final event of experiences in the living world, transforms this reality into eternal verity. Man ascribes to the universe the quality of unity because he sees nature with his mind which is but the exercise of the radiating network of intercellular connections that pervades his being. The Cartesian aphorism, I think therefore I am, we transpose into I think because I am as I am—for we construct our interpretation of nature, the sciences, in the image of our mind. So, Bernard has said that if we had a mind otherwise constructed, mathematics would have another form. [1]

1) Cited from Claude Bernard.

Page 129

The mind of man, as I conceive it, is associated with the entire network of ectoplasmic processes that pervades his whole being and not only with the nervous system, supreme expression in the animal kingdom of nervous structure and function, it too comprising a network of processes, the most highly developed and most completely organized ectoplasm. Mind as the individual's awareness of self and of the world, though hidden and intangible, defying spatial definition, is yet the luminous quality co-existing with life—itself refractory to definition—that which alone traverses all time and all space, though it occupies none, with a speed greater than light which the mind alone can measure. Mind is the function of all those functions of the body, the datum of which is supplied by those activities of the sum-total of the ectoplasms that fix man as individual and as fraction of the universe; it is the glow fed by his being and supported by his atmosphere, the world external to him. Of all these threads man's nervous system is the transcendent expression, the synopsis, whereon hang his history and the history of the universe.

To say mind as a function of all the bodily functions is merely descriptive, not explanatory. Mind, in my judgment, has to be related to some structure. I add, then, that since mind is conceived as generalized, pervasive, it can be related to some structure pervasive in man and general to the animal kingdom. This pervasive structure for me is not the cells, regarded as an ensemble of separate units. I conceive it to be all the ectoplasm, centering in what was the most active region of the egg's ectoplasm, now the nervous system.

Page 130

This is to say that cells minus their surface structure prolonged into filaments—many biologists deny the existence of the filaments, let it be remembered—are a collection of bodies, each of which, though specific both for the organism and for the respective organ and tissue that they compose, and therefore revealing diagnostic characters of no little value, nevertheless lack the very essential ingredient that sets apart life from non-life. It is in this sense that earlier I used the term, quality. For life reveals itself as a bundle of qualitative expressions; no size, no weight is common to it; likewise, no definite number of cells is common to it since it resides in one cell or in billions of cells, from the protozoon just visible under the highest power of the microscope to giant whale animals that are made up of all possible intervening numbers of cells. Neither size nor weight is of taxonomic significance in the classification of protozoa. Within the protozoa, the jelly-fish, the molluscs, or the mammals, as with other groups, we find no size or weight peculiar to the group. Further, there is no size of the cells themselves constant for any group. So too when we take the constituents of protoplasm, water, gases, electrolytes, carbohydrates, lipins and proteins, we find no quantitative determination of any of them common to all life. Two proteins, for example may be composed of exactly the same amino acids in equivalent amounts and yet be different, due to the manner in which the amino acids are linked. Mark you, I am far from saying that what I speak of as the quality of the life-state inheres only in the protoplasmic surface: this is but an extension of the inner

Page 131

protoplasm and hence is in large degree like it. I can never forget this. That is the whole puzzling problem of protoplasmic specificity or of that much debated question of species but a facet of this quality of which I speak? Since as I have said elsewhere I look upon protoplasmic specificity as resident in the ground-substance, the menstruum which suspends the material within the cell, I would not for a moment imply that the cell-surface is alone the life-structure. What I do say is that minus this surface, structured otherwise than that of which it is an extension, there is no tangible display of life-manifestations. And it is to these manifestations as tangible and visible, as phenomena, that we address ourselves in the attempt to approximate an understanding of life—until such time that we can say more aptly what life is. The quantitative determinations so gloriously successful in physical science fail in biology and will continue to fail as long as we continue to overlook, as we have overlooked in vital phenomena, their qualitative aspect so tremendously significant. Yet the vast effort of quantitative studies serves an admirable purpose: these studies tell those who would read how important quality is for the state of being alive.

Mind also, as its primitive precursor, is a quality. And if we relate mind to the nervous system, it holds that the quantity of the substrate nervous system, in normal cases of course, does not determine mind. Brain-weight and brain-size, it scarcely need be pointed out, are not correlated with mental capacity. There is no quantity of mind; there is quality.

Page 132

This mind, I hold, is a product of evolution, its antecedent being present throughout the animal world as expressive of that differentiation in living protoplasm in which inures the very quality of life by maintaining the state of being alive and by implicating this state with the outside world. We have traced back to the simples form of life, the unicellular organism, where this quality is identified in a structure, the peripheralized protoplasm, whose expressions, conditioned by the permanent unicellularity of this organism, are the least diffused and in the lowest terms. With evolution, this structure in multicellular animals was progressively more widely distributed; its expressions concomitantly being less and less of unique location; step by step, the one way mode of the first living thing developed to culminate in the radiating, many branched terms of expression, the human mind.

There are those who hold that mind has not evolved: it could be that it is the sole possession of man. Accepting the doctrine of special creation for each and every living thing, one would certainly include mind in the creation of man. As such, it is a divine gift. Since, as stated at the outset, I write from the standpoint of one who deems the evidence in support of evolution to outweigh the doctrine of special creation, I should be permitted to dismiss any idea of special creation. But dismissing this doctrine as one held by antagonists of the postulate of evolution does not resolve the issue. Many an ardent protagonist of evolution today holds the position maintained by Wallace, who independently of Darwin formulated the theory of natural selection, namely,

Page 133

that whilst the corporeal man has evolved, his soul and mind are divinely implanted in him. The idea of the derivation of the soul from the brute-world is repugnant to many. If man's mind (and soul) is not a divine gift, one of special creation, then it has evolved, even though resident in man only as a new possession. For to say that mind is peculiar to man alone, as, for example, an _emergent_ of evolution, is to say that it has evolved as product of precursor or precursors. As Lloyd Morgan has said: "Two phases of organic development must be distinguished; first, that in which consciousness is either absent or inoperative; ["and secondly, that in which consciousness is a co-operating factor. The first may be termed the merely organic phase; the second, the conscious-organic phase" (p. 262 in _Habit and Instinct_ (1896) by Conwy Lloyd Morgan).] Here I cannot enter into a discussion of the theory of emergent evolution; I reserve this for a forthcoming essay in which I set forth my own theory of the cause of evolution. Suffice it now to say that antecedents are implicit in the term, evolution, meaning the process by which we account for the fact that there is in the living world a gradual ascent from lowest to highest forms of life. An _emergent_ emerges from something; the more striking, new, if you will, the emergent, the more significant as cause is that whence it issues. Vesuvius remained quiescent for centuries before its present eruptive or emergent period. But during its years of passivity the factors for eruption could not have been absent though not operable to the extent of setting up explosions. As I see it, there is no middle ground: either mind has evolved or it is a special creation. And, as has been said earlier, to deny the evolution of the highest human functions is tantamount to denying all evolution. The concept cannot apply to structure alone.

Page 134

That many thinkers, who uphold the concept of evolution, are loathe to extend their support to include a defense of the proposition that mind has evolved is not difficult to understand. This unwillingness is justified by the fact that, although it is believed that mind is bound up with nervous structure and function, the bond is not as sharply put into relief by scientific data as it is wished. And yet good data are at hand. More would be were it not for the physiologists themselves who, in the words of one, himself a physiologist of renown, seem confronted with a mental hazard when they attack problems of the nervous system and especially those of the brain. Mental states, thought for example, considered as a function of the brain, cannot be dealt with, for example, as can the bile, whose excretion is a function of the liver. The brain neither secretes not excretes thought. Measurements applicable to the determination of the manifold functions of the liver would break down if applied to the emotions, the intellect and will. Most of us lack the resolve to speak of thought as a secretion despite the example furnished by two most eminent biologists who could speak of energy as a secretion. One, therefore, perforce, depends upon one's intuition and the hope that the psychologists, who seem to be without the physiologists' mental hazard, will go still farther than the best of them have so admirably gone. For my part, being so strongly an adherent of the concept that structure is for biology inescapable, that vital phenomena are manifestations of structural quality, I cannot abandon my point of view; I have the intuition, and abide in the faith, that we yet establish beyond doubt that mind is a manifestation of structure, part of which is nervous.

Page 135

Some, perhaps, will object to invoking intuition. Yet it has its place in science as has said Henri Poincare, scientist of first rank. As to faith, no other than Claude Bernard has said that when one does not know things and wishes to know them, one makes for them a hypothesis, a belief.[1] More recently, Dr. Compton, in his presidential address to the American Association for the Advancement of Science, has given his estimate of the place of intuition, reason and faith in science. Thus, my faith in hypothesis which I by no means alone hold: mind is a function of structure. But we may never know.

In research, be the subject what it may, we search to find facts and something yet, mysterious, vague, sublime that warms our enthusiasm—that which we never find; that which, having no body, has no name and without which no work of the spirit is ever undertaken on this earth. It is not I who speak thus; I but paraphrase the beautiful language of my dear friend, one of the most charming characters in literature, Monsieur Silvestre Bonnard.

Limiting mind and its rudimentary precursor to the animal kingdom, I range myself opposite to those who endow the inorganic world with consciousness. One argument adduced in support of the idea that consciousness inheres in all nature from atom to man is that the idea preserves the principle of continuity. Note, for example, the following quoted from Sullivan:

1) "Quand on ne connaît pas les choses et qu'on veut les *connaître*, on fait sur elles un hypothèse, une croyance; on suppose les choses comme devant *être* d'une certain *manière*."
["When you don't know things and want to know them, you make a hypothesis, a belief about them; you assume things to be in a certain way." (Google translation)]

Page 136

In *Limitations of Science* (1933) J. W. N. Sullivan says:

> "Although the theory that mind has steadily evolved is perhaps the one that most naturally suggests itself, it is not necessitated by the observed facts. It is quite possible to suppose that sudden breaks have occurred, and that radically new elements have appeared. As we have seen, it has been maintained that consciousness is peculiar to man, and that no animals possess it. Even if we deny this, and insist on attributing consciousness to the higher animals, we might hesitate to extend it all down the scale and to attribute it, say, to the amoeba or to plants. It might be difficult even to stop there, so that presently we should find ourselves regarding the whole inorganic world, also, as conscious. There is, perhaps, no conclusive reason why we should not do this, and there are philosophers who maintain that we should be correct in doing so, but it might be as well, before doing so, to search for alternatives, Any alternative will, of course, suffer from the disadvantage of postulating a breach of continuity."

And later he says:

> "By endowing the atom with a sufficient number of properties we shall make it adequate to explain, not only the phenomena of physics and chemistry, but also living and mental phenomena.
>
> Nevertheless, it seems that the atom will then have undergone a real change. It does not seem that it can perform these new tasks without becoming something different in kind from what it was before. The atom of modern physics, different as it is from the Victorian atom, is different chiefly by being a much more complicated mechanism. It is true that our notions of mechanism have to be considerably extended to accommodate it, but such notions as mass, distance, motion, still remain fundamental in our construction of the atom. From this point of view we can say that

the modern atom is not different <u>in</u> <u>kind</u> from the Victorian atom. But it seems impossible that any extension of this sort of complexity will help in explaining mental phenomena, even if it proves adequate to the phenomena of life. So that amongst the new properties with which we propose to dower the atom, we shall probably have to include a rudimentary form of consciousness. Perhaps there is a hint of this in the modern doctrine that the atom manifests 'something like free-will.'

The chief point in favour of the view we have just sketched is that it preserves the principle of continuity. Scientific men, as a whole, are very strongly in favour of this principle. It may be; however, that their feeling is no more than a prejudice, There are philosophers who believe that the world is a plurality, that it is composed of things essentially distinct. But the principle of continuity has led to so much fruitful work in science that it will probably long remain as a working hypothesis. It will be noted that this way of securing continuity, by postulating some rudimentary form of con-

Page 137

sciousness even for the ultimate particles of matter, involves a sort of dualism. For consciousness is something peculiarly different from the other fundamental properties attributed to matter." [From *Limitations of Science* (1933) by J. W. N. Sullivan.]

Certain of the sentences cited ought to be noted.

If postulating a form of consciousness for the ultimate particles involves a sort of dualism, and if consciousness is taken as something peculiarly different from <u>other</u> fundamental properties of matter, the principle of continuity is badly served by endowing the inorganic world with consciousness. If the principle of continuity cannot be saved other than by breaching it at the point where it begins, it is perhaps best abandoned. That only living things (or even man alone) possess consciousness need not, in my judgment, breach the principle of continuity. For, presumably, the principle is not held to ignore the existence of the living world and the inorganic world. I mean: if the granting that the two worlds differ, permits acceptance of the principle of continuity, it is of no moment how narrowly we limit in the animal kingdom the possession of consciousness. As yet we are ignorant of the existence of a transitional form between life and non-life. While this ignorance persists, life, though compounded in non-life, we set off from the inorganic world. And mind and its primitive forerunner, is, to me, an attribute only of the state of being alive.

One other point deserves comment.

The atom of modern physics, we are told, is a complicated structure. The atom of chemistry, on the other hand, is very much less so. Sullivan himself is of course aware of this difference; he calls it into play more than once with a sort of covert disparagement of the achievements of chemistry in the realm of physical theory. Now, the postulates of science are, as again also Sullivan indicates, always made as parsimoniously as possible. The atom of modern physics, complicated though it is, is not one whit more so than physical theory demands. In the same way the much less complex atom of chemistry is sufficient to meet the needs of the science. If chemistry finds the atom adequate to explain its phenomena, to endow it with other properties would seem to be superfluous. If we arrange the sciences in the order, physics, chemistry, biology—an arrangement justified from several points of view—we may assume on the basis of the greater simplicity of the chemical atom that the biological unit of structure is more simple yet. Since an atom simpler than the physical atom satisfies the needs of chemical theory, it may be that a simpler atom than the chemical one suffices for biology, that is, for interpreting living and mental phenomena. As actual experience goes, the atom of biology is far less complicated than that of chemistry, and is even less so than the atom of physics, for the simple reason that in biology there is no atom: its unit of structure is the cell or other form of protoplasmic system. For the vast number of problems peculiar to biology, the chemists' concept of the atom can be ignored, and the physicists' atom need not ever be invoked. If, as Sullivan earlier tells us, "it is the electron that is the key to the universe" not only biology but also chemistry lies outside the universe as viewed by physics. Then here certainly the principle of continuity is more than breached; it is rendered nil: the universe thus becomes at least triune—physical, chemical and biological. Either this, or a change in the concepts of physics [is in order].

Page 139

That these concepts are in such a state that the picture of physics lacks definition is generally known. It is a state that presages change, revolution even, in our ideas. Since it cannot be foretold whence will come the change or whither it will lead, the making of final statements, as if ultimate truth were won, ought to be always avoided. It may be that "one of the real tasks of science at present is to deduce the laws that govern matter in bulk from the laws that govern its ultimate constituents", and that "the reduction cannot be effected the other way round", as Sullivan tells us in speaking of the electron as the key to the universe. But <u>cannot</u> is a strong word; it is not impossible that a new concept be forthcoming through reduction the other way around. However, I, for my part, doubt that the revolution will come by means of any simple shifting of emphasis given to the meaning of mechanism whose break-down is in large measure held responsible for the present situation in physics. Professor Whitehead, for example, asks "What is the sense of talking about a mechanical explanation when you do not know what you mean by mechanics?" [Alfred North Whitehead (1925) *Science and the Modern World*, p. 378] Also he tells us: "The only way of mitigating mechanism is by discovering that it is not mechanism" [Whitehead 1925: 76]. Then straight way he elaborates his theory of organic mechanism ["theory of *organic mechanism*" (Whitehead 1925: 80)], very much as one would say that Homer did not write the Iliad but another man by the same name did; whilst denouncing mechanism, he retains both in the term and in the concept, organic mechanism, the odious patronymic, so to say.

The far-going abstraction, the reduction of matter to the point of its non-existence, leaving in a colorless, soundless, odorless, tasteless universe (expanding or contracting?) only a penultimate if not an ultimate abstraction, the electron, which, as light, is wave-like or corpuscular or both; the uncertainty principle which by being denominated by some as indeterminism allows the interpretation that the atom possesses something akin to free-will, and that cause and effect are no more; the quantum theory and the relativity theory—in short, all the brilliant achievements of modern physics made possible by quantitative methods of superlative elegance by their wealth embarrass the science; it is surfeited with concepts not yet clearly defined. If this picture of modern physics be true, it is to be doubted that its concepts can have any but very limited applications to the problem of vital phenomena. The very quality of the state of being alive proscribes the utilization of physical concepts even when these are most clearly acknowledged and accepted by all physicists without due recognition of the concrete substantiality of the life-state so far removed from the product of abstraction, the electron. I doubt that we can arrive at a concept for living substance by an extension of the methods of abstraction employed in the investigation of the "ultimate constituents" of matter. For a living thing, the structure where life resides is the ultimate constituent. Seeing no immediate chance of eliminating structure from biological concepts, I defend the thesis that the mechanistic concept of physics, even where its status in physics is unquestioned, does not apply to structures manifesting vital activities.

Life, then, as has been stated, by virtue of its peculiar organization of inorganic constituents that are amenable to quantitative analysis, which, however, is incompatible with the life-state, is qualitatively apart from non-life. In the sense that the pattern of the life-state is a composition of universally occurring elements, a structure of the most common of these, that probably arose from these, depends irrefutably from and on and returns to these, and evolves from simplest to highest structures—it gives us a picture of continuity, of unity (with them). Nor the living thing is, as it were, the simulacrum of the universe. The hypothesis that limits consciousness, and its forerunner, to living things is no more a breach of the principle of continuity than is life itself. And to deny altogether the existence of consciousness, as some do, would not save the principle, whilst life remains.

Having no singly localized substratum, mind has no spatial definition. It knows no <u>here</u> and <u>there</u>. Having no moment, it encompasses all time. It recognizes no <u>now</u> and <u>then</u>. Space and time have limits because beyond these are yet unexplored regions of mind. We create nothing only find out what there is. When we have reached finality, when we have exhausted all mind, the universe will come to its end. For mind is the mirror of eternity wherein we see in part the image of all nature, past, present and what is yet to be. And when this image floods full the glass, so that we see stark truth in its entirety, the sun of our soul will be no more in our shrunken, shriveled universe.

The activities of man's mind express themselves in various forms of spiritual endeavor; art, science, religion, ethics. But mind, not being an entity, in the meaning of a concrete phenomenon, these modalities of expression as depositions of its diffuseness are never quite pure but intermingled with each other. There is no art wholly free from science, just as there is no

Page 142

real science that lacks the quality of art. Also, science and art are alike inasmuch as in each are forms that represent a fusion of mind with handicraft, music and mathematics not of necessity employing the hands. They are unlike in the degree in which they make appeal to the emotion and to the intellect. But aesthetics, even music, is never completely divorced from intellect. Mozart, whose inimitable music encroaches upon the divine, in one of his puckish moments, moments not rare to him, was able to formulate waltz measures which one can arrange beautifully in any order set by the fall of dice. Music has no less emotional appeal because it has laws. Whereas to me much of Bach seems stilted, perhaps because of his use of almost mathematical laws of composition, to others he is ever the great spirit of music. There is a science of harmony and I sometimes think that its greater comprehension by investigators in the natural sciences would mean much for the advancement of our knowledge of nature. He who views the mural, <u>The Last Supper</u>, glorious still in its fading beauty, does not think of the artist as a skilled anatomist acquainted with the fact of the circulation of the blood for which his fellow-countryman, Ceisalpino, receives credit and less of the perfect geometrical design of the painting, worthy of the engineer that da Vinci was.

So too science without feeling lacks power of transformation. In its very beginning, all science derived from feeling. The great scientists without exception have been men of powerful imagination, stimulated by a feeling for beauty. He who brings us one step nearer to truth moves in his quest under the impulse of sentiment. You may not wholly subscribe to Claude Bernard's

Page 143

saying that reason and reasoning alone are the source of all our error, sentiment being a surer guide; but you will agree that, without sentiment, scientific endeavor lacks initiative.
[Claude Bernard 1927, 1865 in French]

The botanist, Osterhout, speaking of his predecessor [Jacques Loeb] at the Rockefeller Institute for Medical Research, has said:

> "If we realize that the great driving force of his life lay not only in a powerful intellectual urge, but also in a profound emotion we may better understand his zeal and why he attacked most eagerly the subjects where mysticism was most strongly entrenched. No matter how great the difficulty, he seemed determined, as far as possible, to reduce everything to mechanism and his courage was often justified by startling success. When unable to solve the problem his keen hypotheses, often startling in their audacity and beauty, and attractive for their simplicity and clarity, aroused and stimulated his readers. Often his dreams were as inspiring as his actual discoveries." [W. J. V. Osterhout, (15 September 1928). "Jacque Loeb," *Journal of General Physiology*, vol. viii, no. 1, pp. ix-xcii, p. ix-xii]

Thus, even a false motive by its emotional power may stimulate scientific endeavor.

Opposite these witnesses—and there are others, the master-builders of science among them—are, to be sure, such who maintain another point of view, as for example a biologist ranked today as among the foremost in his science has written:

> "This pageant makes an irresistible appeal to the emotional and artistic sides of our nature. Hence not

without a feeling of jealous regret, the old-fashioned embryologist sees these gems of nature consigned to test tubes for chemical analyses, to centrifuges to disturb their arrangements, to microdissecting instruments to pick them to pieces, and to endless tortures by alterations in the environment to disturb the orderly, normal course of events. For, it is the automatic self-contained perfection of the developmental process that holds our interest. Yet we feel, too, that if the mystery that surrounds the study of embryology is ever to come within our comprehension, we must try not to be sentimental and have recourse to other means than description of the passing show.

Page 144

> The recompense, we hope, will be to substitute a more intelligent interest in place of the older emotional response to the order of nature." [Thomas Hunt Morgan. (1927). *Experimental Embryology*, New York: Columbia University Press, p. vii.]

This, I deem, is the statement of one who is sure neither of himself nor of his science and thus hypothecates much too much for both.

Thus, although art and science are two different modes of endeavor, they do not entirely exclude one the other. Though the proportions of the emotion and of the intellect infused differ, each has a basis in sentiment or feeling. Both employ the corporeal being as a vehicle of expression; except for unrecorded music preserved only in memory of a folk, folk-lore, as Homer's poetry, and mathematics, which is not entirely dependent upon use of the hands, art and science are truly maneuvers, literally, in employing the hands and figuratively as mind's stratagems.

The foundation of religion is faith, its essential quality resting in a belief from which it never departs. It is a purely spiritual manifestation, without body or the use of body, projecting itself into the future, the soul's inquiry after first and final causes. Thus, as do art and science, religion has its roots in feeling, although the feeling is more vague. Belief preceded reason. Man was religious before he was philosophical as he was philosophical before he was scientific; before he had a scientific interpretation of nature, he had fears and superstitions whence arose his interpretations of life and of natural phenomena. Stripped of fears and superstitions, his religion, be it belief in God, in nature or in humanity, remains an imperative need that still has power to lead the world. However much we intermingle religion with art,

Page 145

science and philosophy, religion stands apart from these as the soul's aspiration to penetrate the mystery of being by faith.

This is far from denying that religion has been an influence in art and science. On the contrary, art was at its highest in the springtide of religious fervor. We shall probably never see another cathedral of Chartres. Instead, we have church edifices that one has aptly named silo-churches. After Mozart's swan-song, his <u>Requiem</u>, came "the pulse of the machine", with lust for gain in the podium, moving from the overture of the industrial revolution to the present crescendo of might.

Modern science also came first out of religious feeling. Underlying all science is a <u>belief</u>, a belief in an order in nature. And this sprang from the idea of God. Speaking of "the greatest contribution of medievalism to the formation of the scientific movement", meaning what he calls "the inexpugnable belief that every detailed occurrence can be correlated with its antecedents in a perfectly definite manner, exemplifying general principles", Whitehead goes on to say:

> "Without this belief the incredible labours of scientists would be without hope. It is this instinctive conviction, vividly poised before the imagination, which is the motive power of research; that there is a secret, a secret which can be unveiled. How has this conviction been so vividly implanted in the European mind?
>
> When we compare this tone of thought in Europe with the attitude of other civilizations when left to themselves, there seems but one source for its origin. It must come from the medieval insistence on the rationality of God, conceived as with the personal energy of Jehovah and with the rationality of a Greek philosopher. Every detail was supervised and ordered: the search into nature could only result in the

vindication of the faith in rationality. Remember that I am not talking of the explicit beliefs of a few individuals. What I mean is the impression the European mind arising from the unquestioned faith of centuries. By this I mean the instinctive tone of thought and not a mere creed of words.... My explanation is that faith in the possibility of science, generated antecedently to the development of modern scientific theory, is an unconscious derivative from medieval theology." [Alfred North Whitehead, *Science and the Modern World* (Lowell Lectures, 1925). New York: Macmillan; Free Press, 1967: 12.]

Page 146

The very basic principle of science is intuitive. It is also primitive in its source, for the history of the development of science exemplifies the God-idea transferred to nature. Some of the most agnostic and the most atheistic scientists have not succeeded fully in escaping the sub-conscious idea of God.

Morality, ethical behavior in the broadest sense, is conduct, and conduct means action, premeditated or not, conscious and unconscious. Action in ethics differs from that which portrays art and that which builds the foundations for science. Ethical behavior involves the whole corporeal being. Codified, it is the collective experiences of the species, hence it resembles all other biological problems that have an historical aspect always to be considered. Although also a spiritual manifestation like religion, it is less a belief than the shadow of a memory; it lies less in tomorrow's expected sun and more in the indistinct haze of vanished yesterdays. In other respects, it differs from the remaining spiritual manifestations: art endeavors to reveal relationship, science expresses relationships, and religion is a projected relationship. Ethics is relationship. The problem of ethics thus lies in that domain of biology whose subject matter is the relation to the surroundings that is necessary for the organism's continued existence and perpetuation of its kind.

Ethics thus for man is inescapable. It was not acquired by man or born in him. Man was born in ethics, for what is ethics for man is, in its primitive form, that which preserves and sustains every animal in its relation to the environment. Structure has not only survived, but has evolved. What is responsible for survival and evolution is the interdependence of organism and environment. It may be said that ethics (or morals) gives us the rules to be followed in order to attain the good and to avoid the evil. I say that nature makes it mandatory that man achieve the good and escape the evil; ethics is the supreme gift to us from the nature of the good, the beautiful and the true. It is thus less a body of rules than a mandate springing from the very initial impulse whence came first life, through which alone life can persist.

The sub-division of the fields of knowledge is apt often to be arbitrary, and doubtless it will remain so as long as knowledge is imperfect. There are regions in the most exact domain of inquiry, the physical sciences, that defy strict localization, as, for example, the term, physical chemistry, implies. The subjects, biological chemistry and biophysics, indicate also that fields may have vaguely defined limits. Some simplest living things are classified as both animals and plants. In fact, biology abounds in hazy boundaries, as one of its basic problems, that of species, in its unclear state, shows. There exist borderlines only in exact science.

In what are denominated the social (or spiritual) sciences, domains are less clearly marked off than in the natural sciences. There are problems considered as belonging to the province of economics that could justly be held as sociological and vice versa. Often it is difficult to distinguish between one or the other and history.

Page 148

Uncertain markings of the terrain of the social (spiritual) sciences leads, unfortunately, to a great deal of charlatanism, to the detriment of the great work of serious and devoted students: too many Cagliostros have hearing. Politics masked as political science may kill the solution to a purely economic problem to the detriment of the consumer, who is a taxpayer as well, to satisfy him as voter. The vague boundaries among these sciences are again indicated by the strong influence that one of them may exert on the other. Thus, an apparently perfect theory of money is not uninfluenced by psychology, and religion even, although religion be far removed from the money-market. In our all too recent memory, the moneychangers have been, allegedly, driven from the temple with almost immediately—ensuing dire consequences—to the temple.

Realizing that specialization is a means to an end, one ought to avoid the drawing of too hard and too fast lines, a danger arising from specialization. Especially is this true when one invokes the principle of the division of labor, a principle badly employed even where it has legitimate play. And it has less play in speculative thinking than in exact, manipulative activity. As Claude Bernard has said:

> "Science is one as the spirit but the division of labor is necessary for execution. Therefore, the specialty ought never be in the head or speculative part; it ought be found only in the hand or active part."
> [Claude Bernard 1927 (1865 in French)]

Specialization is a device for work; the spirit remains single, having various manifestations vaguely defined.

Page 149

Ethics is here envisaged, in the light of the fore-going, as a separate manifestation, as are art, science and religion. Although it is an art, in the sense of being the practice of a way of living, and a science in so far as it endeavors to delineate the principle or principles of conduct, ethics, by virtue of peculiar differentia, may be set off from both art and science as a distinct spiritual display. Although earlier incorporated with religion, ethics came to be set apart from it, as somewhat later it became separated from metaphysics. Some of the greatest formulations of ethics derived from founders and exponents of religion; the founder of Christianity was one whose ethical teachings are still widely embraced. Similarly, men whose names are great in the history of metaphysics have made the largest contributions to ethical theory. However, with sub-division of the domain of human inquiry, ethics came to be divorced by many from religion and from metaphysics even.

Page 150

Chapter 8 - **FEELING, REASON, AND WILL** – pp. 150-176

Relationship is the key to understanding nature. As far as the mind can encompass the universe, it beholds a vast interlocked organization of planets and of stars, a gigantic combination of parts in relation. This relation is one of dependence, a dependence of product to source. And when the mind in its greatest venture of temerity moves out from the known to the unknown, and endeavors to traverse space and time, it comes to a region too cold and blank for thought. There lies the unpenetrable mystery of first cause that we may never know. Futile the voyage that extends therein.

But we can know the earth and our immediate universe, though we see it as yet only in part. These too exemplify the relations that science endeavors to understand. Here also the relation is one of dependence. This earth depends upon the sun whence it came; its dependence is that of product to source. Nature on this earth likewise reveals to us no independent units; nature is a vast harmony of parts relevant to each other. It has been said that the environment fits the living organism and that the living organism fits the environment. The truth rather is that organisms depend upon the environment. This too is a dependent relationship of product to source. For, although we do not know, may never know, that life did not come to earth borne as infinitely small particles from another planet or that the first form of life as an act of the Creator, which since has evolved, or that

Page 151

life arose as the result of some combination of inorganic materials present in the then-existing world to form organic compounds and not of products of life-processes—whence evolved life, we can hazard only a speculation, based on what evidence is available. In this wise we suggest that life came from non-life.

The strongest evidence for this suggestion lies in the indispensable and persistent dependence of living things upon the inorganic world, upon gases, electrolytes and water. These alone constitute the need common to all living things, animals or plants, whatever their place on the scale of evolution. Man is as dependent on oxygen, mineral salts and water as the lowliest form of life. Some lowly forms of life can maintain themselves exclusively on inorganic compounds. Thus, the postulate is plausible that life came from an inorganic medium. But, I repeat, we do not know.

As the mind surveys the course of evolution, it encounters at every critical phase a gap remaining. How life arose is one such gap, as are the rise of Metazoa, the origin of vertebrates from invertebrates, and the transformation of man from some ape-like progenitor. Although these gaps may never be filled, the mind is intrigued to offer tentative suggestions consistent with the now-known facts. At any rate, these lacunae are not sufficient to rule out the hypothesis that evolution has taken place in the living world as in the non-living. Moreover, the situation could be worse but for the existence of amphibia, which represent a link between wholly aquatic and guilded animals and wholly terrestrial animals with lungs—and but for the finding that connects reptiles and birds.

Page 152

If, however, you object that the gap between life and non-life is so much greater than these others that it cannot be considered as belonging in the same category with these, however plausible in view of evidence of the inescapable dependence of life upon the inorganic medium, let us consider the problem of dependence from another angle.

Consider the entire sub-kingdom of multicellular animals quite apart from any hypothesis concerning its origin from unicellular life forms, as if these latter did not exist. The one and only feature common to these, from sponges to man, is their development from eggs. No multicellular animal exists that does not develop from an egg. The egg then is source, the adult being in a sense a projected egg—in a sense, because the egg depends irrefutably upon its inorganic environment. I do not mean that <u>all</u> animal eggs depend <u>only</u> on the inorganic environment, for this would exclude the eggs of mammals and of such forms whose eggs cannot develop outside the maternal body. I do mean that that which alone is common among them is the inescapable dependence of the egg's development upon the inorganic medium. Beginning with the egg at its initial phase of its process of development to the adult stage, we thus take the egg as the given source. Hence the inorganic medium is not the source, and the dependence is not one of product to source. This is obvious. But just as obvious is the fact that the egg <u>as</u> <u>developing</u> <u>system</u> is not the entire source of development. For we have yet to encounter development apart from the inorganic environment. We can think of such, of course; but in experience, do not meet it. A theory of embryogenesis that denies

Page 153

the indispensable play of forces in the inorganic environment of the animal egg avoids the very problem involved. Given how the substrate egg successively sundering itself into more and more minute chambers maintaining interconnections; how a system emerges that is capable of self-maintenance through the successive appearance of new relations of its partitioned parts to the external medium; and how the ontogenetic process is one of progressive dependence upon extrinsic factors—even the most thorough-going proponent of the preformation-doctrine must admit that embryogenesis is dependent on the environment.

Understand, please, that in citing the egg as the one possession of multicellular animals, I do not abandon my position with respect to the postulated rise of life from non-life, from a non-living environment existing anterior to the appearance of life. I am only willing to allow that, if this postulate relates in your judgment too far in the remote past to that which we can never know, and in view of the hiatus of the evolutionary scheme between unicellular and multicellular beings, my argument restricted to multicellular animals considered, apart from any hypothetical descent, as _ab initio_, warrants the conclusion that the inorganic environment is a dependent source of life and life-processes. Considered as an event, or rather as a series of events, an embryogenesis is self-sufficient, self-maintaining and self-regulating only in terms of the indispensable dependence upon factors in the environment. The egg in its own efficient cause of development—but this reveals itself, moves on to effect, only within the medium of its surroundings.

Page 154

Now, of course we reckon with forms of dependence other than that upon the inorganic milieu. Viewed in the large, the animal kingdom depends for its food upon plants. The isolated instances reported of animals that are capable of maintaining themselves on inorganic compounds alone may be overlooked. Also, carnivores depend for their food indirectly upon plants. And since our discussion concerns itself with animals, the dependence of one species on another or upon plants for food must be considered as being necessary during a large span of the individual's life history for existence. The difference in dependence upon inorganic and organic substances is one of degree only.

This leads me to say that we ought not to substitute the word interdependence for dependence used in the foregoing. Organisms depend upon the inorganic environment but not this environment upon organisms. The world of water, minerals and salts and gases is independent of animals and plants. In animal nutrition the case is the same: animals depend upon plants but not necessarily plants upon animals—taking both kingdoms in the large. It is true that plants utilize carbon dioxide and nitrogen compounds excreted by animals, but this does not mean that animal-life is necessary for the maintenance of plants; the plant kingdom can exist without animals, deriving needed sustenance from the sun, the atmosphere and the soil that contains plant, but not animal, residues. There can be no question here as concerns the sun, water, salts and gases. The nitrogen cycle, for example, could run from the starting point of free nitrogen,

Page 155

by way of either nitrogen-fixing plants (bacteria) or of the root-nodules on the very common leguminous plants, to plant protein; from plant protein to plant residues in the soil; from these residues to ammonia by bacteria back to plants. In other words, the dependence between animals and plants, not being necessarily reciprocal, is not properly an interdependence.

The relation of animals to plants, of carnivores to herbivores, and of parasites to hosts dictated by food-requirements can in no wise be defined as interdependence. These are modes of dependence of the eaters on the eaten. Only the first named can be regarded as general to the entire animal kingdom, the others being sporadic and occasional, do not allow a basis for general application.

In addition, instances of commensualism and symbiosis cited to show interdependence are not strictly speaking such. At Roscoff (France) is the flatworm, Convoluta roscoffiensis, which takes in green algae and is said therefore to "carry on a partnership in nutrition, as it were". Whereas the worm ceases to grow without the green cells or when in the dark, the algae can lead an independent existence. Thus the dependence is not strictly mutual; the interdependence is limited and not equally indispensable.

Quite apart, then, from the point already made, namely, that mutual aid considered as a factor in the origin of species ought to refer only to members of the same species, is this, led up to by the fore-going discussion: symbiosis or commensualism are not modes of a mutual aid, between species of the same or of different kingdoms, unless the symbionts or the commensals render and receive in something like equivalent value.

Page 156

It should also be noted that the term, co-operation, applied to any of these examples cited is incorrect. Co-operation means a joint effort toward the same end. We have no knowledge of an identity of ends as the basis in the living together of green cells and a flatworm, or in that of any symbiotic or commensal organisms, even where the life association is equally indispensable to each.

My first point, then, is that in the animal world the maintenance of individual life is in a wholly dependent relationship with the environment, meaning first the inorganic, and second the organic and living, environment. My next is like unto the first: the maintenance of the continuity of the species also depends upon the environment.

As we have seen, no animal egg develops apart from its medium. Include now the asexual mode of reproduction sometimes found alternating with the sexual in certain worms and some other forms lower in animal scale, together with protozoa, animals that have no sex and therefore no eggs. Then we alter our statement to read that, since no reproduction is possible apart from the medium, reproduction depends upon the environment. Here, we can extend the scope of environment to include the living, reproducing thing itself because all forms of reproduction, sexual by way of fertilization or parthenogenesis, and the various modes of asexual reproduction, is but a sub-division of that which reproduces itself. Reproduction means just that, as we have seen—the sundering of the thing which differentiates itself to become like that of which it was a part. Then not only single location in some kind of environment— for, the majority

Page 157

of the members of the animal kingdom require a non-living, inorganic environment, which is absolutely indispensable for the process of reproduction, whatever its form—but also the living environment, the reproducing thing itself, is indispensable. We know of no reproductive process apart from location in space either of the external medium or of the reproducing thing.

If we admit the fact of the dependent relationship of organism to environment, then it follows that this dominates all other communications. Indeed, all relations of organisms, all their activities, derive from the fundamental dependence. Since the time of Francis Bacon, thinkers, natural scientists and moralists alike, have often voiced in one way or another the idea, true for animate nature, that living things reveal two kinds of function around which center all life-activities: nutrition and reproduction. We commonly speak of functions apart from reproduction, as "vegetative" and "animal", designating nutrition by the former and all else by the latter; these refer to the animal's activities as an individual. Taken together, the entire complex of the individual's behaviors is for the species vegetative and only those concerned with reproduction are not; individual life is subservient to the continuity of the species in which alone is the possibility of progress of evolution.

The law of environmental dependence, for this designation is justified, dictates, as was said above, the single location of organisms in space. Let me hasten to add that this does not mean that all members of a species are restricted to one area alone. Species differ widely with respect to their geographical distribution; many are cosmopolitan and some of them vary in

Page 158

many interesting and puzzling ways which we do not yet understand; the incidence of parasitism as shown in Professor Peréz's remarkable studies is a case in point; Professor Vandel's researches on geographical parthenogenesis is another; less well known are such reported instances as Miss Ida M. Hyde's findings on differences in mode of gastrulation and the variability in the pigment in eggs of the edible European sea-urchin, although this is not strict since some animals at Naples produce eggs with and others without the pigment-band, but at Roscoff (France) no eggs show it. Nor is it meant that a single species pre-empts a locality. The requirements of each species of animal being peculiar, the various species tap the vast environmental reservoir in different quantitative and, to some extent, different qualitative degrees. Hence organisms are distributed in space according to the availability of needed supplies. However much the climate, food-supply and prevalence of enemies may vary from zone to zone occupied, whether permanently or sporadically, by an animal species, this is single in the sense of having, despite these differences, the property of satisfying the species' needs. This common property constitutes for the species the singularity of location. Moreover, the differences in geographical zones supporting life are not so great after all, when we think of them for a moment. Take, for instance, the extremes of temperature between which animals can live: this is relatively a short range. It is much the same with such factors as salinity, hydrostatic pressure, barometric pressure, etc.

Page 159

The law of environmental dependence, by dominating all relations between organism and environment, dictates the relations among members of the same species. These relations are related to each other by virtue of having a dependence on their environment. In this respect they are one. True, within a species, varieties presenting easily discernable structural differences are often encountered, but there is nothing to indicate that the needs of these varieties extend beyond the limits of the species any more than the structural variations do. Structural variations within a species may be less pronounced than the sexual dimorphism or than structural changes that take place at the breeding season, some of which in either case are so great that animals of one species exhibiting them were long classified as distinct species: for example, many marine worms, as the syllids and nereids. The difference between worker-bee and queen is due to a difference in food, but once attained, does not imply changes in structure or in needs that place either outside the species. At any rate, the worker-bee, a sterile female, lacking the capacity to perpetuate the species, is outside the law as regards reproduction. With respect to the drone, as of all male members of animal species in which sex is set up by the males—in elaborating male and female producing sex cells in equal number—it may be said, in the light of the known facts of natural and experimental parthenogenesis, that one great function of this is to avoid catastrophe for the species: the production of males only. What a human race would there be, and how long could it persist, if this be true for human males also, if for mankind likewise males were needed to avoid a human world made up of males only!

Page 160

Members of a species, be they humans, mackerels, earthworms or amoebas, are individuals exhibiting individual characteristics in more or less striking degree, depending upon our knowledge of them. Members of a species are not of a standardized form, weight and size even when they are of the same age, are subjected to the same physical factors, or take in the same food, etc. We cannot, perhaps never can, write a chemical formula or give a mathematical expression to cover the members of a species. Nevertheless, we can say—and this is true despite the deplorable state of the whole species-problem in biology—that the members of a species have a common bond that relates them to each other. This is the identity of their dependence upon the environment. Although a named locality teems with representatives of many distinct species, the needs of each set each apart from the others and make of each (species) a unit.

Finally, the law of environmental dependence entails that we regard equilibrium in its proper light. Let us repeat the words of Claude Bernard quoted above when it was stated that it is necessary to disengage the word, equilibrium, from the meaning of the term, struggle for existence:

> "Without doubt herbivorous animals
> nourish themselves of plants and the
> carnivoures of the herbivores. ...
> These results which assure the cosmic
> equilibrium are the consequences, as
> we shall show later, of the general law
> of the struggle for existence according
> to which nature can engender life only
> by death, creation by destruction. ...
> That which is observed in the intimate
> phenomena of nutrition, in the depths
> of our tissues, is manifested in the
> grand cosmic phenomena of nature.
> Living beings can exist only with the

> materials of other beings dead before them or destroyed by them. Such is the law." [Perhaps from *An Introduction to the Study of Experimental Medicine* (1927 [1865]) by Claude Bernard.]

No, Claude Barnard was much too clear a mind to give false meanings to terms. He is the greatest physiologist, not only because of his achievements in experiment, but also because of his straight thinking. Nevertheless, this passage might be construed

Page 161

as one setting up a wrong conception of equilibrium. If a carnivorous animal eats an herbivorous one, which in turn has utilized plants as foods, there results no equilibrium. There has taken place a release of energy from the sun and other external factors to the plants, from these to the herbivore, and from the herbivore to the carnivore. There has been no exchange but only a progressive surrender of energy during the stages of which all except the initial and the terminal forms disappear. The plant and the herbivore having disappeared, there is no problem here comparable to that of the lady and the tiger, à la Stockton; we are not left to guess what has happened; it is merely the picture of big and little flea ad infinitum in shorter terms. Two terms having been successively engulfed, we may ask what is in equilibrium with what? If we extend the problem to embrace not one member of each term but as many as you wish, the result is the same throughout the series—it assures no balancing, but only a destruction, a disequilibrium. Let us say that there are plants sufficient in quantity to satisfy a given number of hungry herbivores, so that a balance obtains between eaters and the eaten. The moment that the food is taken, the balance obviously disappears. To say that nature engenders life only by death is only to introduce an element of confusion. Life in the examples given is sustained by death (the death of other lives); but life is engendered only by life—be the term, life, general or applied to a named species, especially if it applies to a named species. True, plants utilized the products of animal bodies but this is evidence of cyclical change and of conservation, not of equilibrium. Finally, that which occurs in our tissues cannot be likened to the grand cosmic phenomena

Page 162

(of equilibrium), though living things exist only on other living things, for the tissues sustain life not by maintaining an equilibrium, but by postponing it. A system, be it a living or an inorganic system, in equilibrium is dead. To qualify the term, equilibrium, as some do, by the words static and dynamic, means nothing. Equilibrium either is or is not; if it is, it is static. A system said to be in "dynamic equilibrium" is one in a state of not-equilibrium. Life is incompatible with equilibrium; the terms, life and equilibrium, are antipodes.

Life can exist only in strictest accordance with the law of environmental dependence. Every living thing, while its life lasts, is the receiver of continuous streams from its inorganic environment, and by far the majority must receive these from the organic environment as well. If, as is held, all reactions in chemistry are reversible, the apparent irreversibility being due to the limitation of our manipulation, then life is not a reaction since it runs in one direction only. When the living thing ceases to be able to receive from the environment, it is no longer alive. Death is the point of reversal.

It follows from all this that every organism, animal or plant, is to be regarded as a definite entity that has independent value, existing for its own and for its species' sake and quite apart from its importance to the well-being or the ill of other animals and plants. Said otherwise, each species has dependence upon its environment independently of all others. This relation is determined for each species by its structure. Members of a species take in from their environment according to the protoplasmic organization peculiar to the species. All inflow

Page 163

toward the organism, be it food, water, gases, or other, is determined by this organization, which regulates the quality and quantity of intake. The play of physical and chemical factors set up changes in the milieu to which the protoplasmic organization responds. What this structure is and how it responds, these are questions, which, if they can be answered, throw light on the problem, the state of being alive.

As we have seen, protoplasmic structure is common to all living things. So far as we know, life does not exist apart from this structure. What this structure is, it must be admitted, biologists have yet to agree. A body of fact, however, indicates that agreement can be reached. Without introducing here the evidence, we may affirm that it permits us to define the living protoplasm par excellence as that part of the protoplasmic system that is capable of putting into play, without abatement, all the features and activities characteristic of the life-state of the given system. Experiments on animal eggs prove that the living protoplasm is the clear matrix of cytoplasm, with the surface structure plus the nucleus of the egg, or that of the spermatozoon. A minute speck of this substance, though it is a small fraction of the whole egg minus the egg-nucleus, develops as the whole egg if fertilized. A fragment containing the egg-nucleus responds to experimental means with parthenogenetic development.

What alone is common to protoplasmic systems, be they plant or animal, is the living ground-substance peripheralized; for, as we have seen in an earlier chapter, although a discrete nucleus may be absent, a bounding surface never is. Further, this surface is visibly differentiated from the inner

Page 164

mass of ground-substance. To this peripheralized, as <u>sine</u> <u>qua</u> <u>non</u> of protoplasmic organization, one can relate the responses of the entire organization to changed configurations in its external medium. Having done this, further exposition would be otiose. (See also my book ...) It is sufficient to repeat that life-processes reveal themselves as activities displayed along the protoplasmic limiting surface.

Common to all living things are: dependence on the environment, protoplasmic organization, and structurally differentiated living surface-structure, where displays of life-processes are visible. Each species may be regarded as a unit in that it comprises an aggregate of individuals having like dependence, and similar structure and behavior. Every species is a closed circle with respect to every other, with only the environment touching it at all points—a circle that is never squared and hence never in comparable contact with other species, which are themselves circles.

One such activity is the reporting of the ceaseless changes taking place in the environment, the surface being the one and only region of the living protoplasm sensitive to these changes. This capacity or aptitude to receive impressions, or this faculty to put to proof, that is, to exhibit response, may be designated as sentiment or sensibility, respectively. This property, I repeat, is common to all animals and, in its grades of display above that by the simplest forms, sharply separates animals from plants. Its highest expression is man's nervous activity, his spirit, soul and mind.

This sentiment, or sensibility, is in the animal kingdom

Page 165

the primitive foundation of nervous activity. We have earlier traced the evolution of nervous activity to its source, namely the primordial and simple reaction of the cell-surface to the environment thence, as it becomes more closely bound to nerve cells, to nerve net, to nerve thread, and to central nervous systems of vertebrates and of man, these being all ectoplasmic structures. And never is this property of sentiment or sensibility lost; it is not only primitive, it is also persistent, being man's heritage from his primordial ancestor, the lowest animal form. More, in his individual development it appears as the first nervous activity, long before reason and will. We feel before we reason or will. In this sense Claude Bernard is correct in saying that feeling is a surer guide than reason.

Montaigne's query, Que sais-je?, or Tennyson's cry, "Behold we know not anything", has again and again echoed through the ages. For we be but children crying in the dark. What we know, if we know anything, begins with our experiences with the outside world and centers in what we feel, our instinct and our intuition.

A theory postulating the natural origin of ethics has to be formulated in consonance with the following:

1. It must recognize that the law of environmental dependence, which holds for every living thing, dictates that every species, independently of all others, depends on its outside world. I use the word, must, which I almost never use, because life is not possible save in this dependent relation.

2. It must reckon with the fact that protoplasmic structure, the sole structural possession of all living things, is characteristic for each animal species, and with the further fact that,

Page 166

at the differentiated surface alone, the one region found in all protoplasmic systems, vital reactions reveal themselves as expressions of the species-specific protoplasm. Again, justifiably, the word, must.

Examine now these statements.

Members of any species are more closely related to each other than to any outside the species by virtue of similar, even identical, dependence on the environment. Human morality is that intangible intra-specific bond arising out of common dependence; for mankind, a single species, is no less dependent upon his outside world than any other animal species. Morality, defined as fellowship, fraternity, and equality, embodying whatever imperatives of justice, equity and of duty apply to the whole of humanity. No lesser application of morality is admissible. Any definition of human ethics restricted to a social group as clan or nation or even one embracing all nations as such is inadmissible. Clans, states and nations are biologically artificial, although they have been beneficial to human progress. Such human groups have only occasional counterparts in the animal kingdom. Where such are found, as has been sufficiently shown, they arise later than the species-group; they are not common to animal life as is the species group; and where they appear, as among social animals, they are adventitious and sporadic, neither they nor any fore-runners paralleling the course of evolution. Nor, yet, are family-life and family-ties primordial of human morality; these were late in manifesting themselves, and are within the domain of the aboriginal source of morality, the intra-specific bond. Where such family relations exist, they derive from parental care

Page 167

and filial dependence, themselves referable to a late-coming consequence to sexual reproduction, a species-diagnostic, and a mode of reproduction of later appearance than the asexual.

Thus, I do not hold with any system of ethics predicated upon circumstances in time or space. A human morality to be such ought apply to all mankind whenever or wherever found. A morality whose definition "would perhaps require some limitation on account of *political ethics*" (italics mine), according to Darwin; or one that is less than "the *Law of Honour* (italics Darwin's), that is, the law of the opinion of our equals, and not of all our countrymen", is not my conception of a morality whose origin is in animal life. Whatever the time or place, however large or small the human group, with this or that moral code, its ethics, if postulated as of natural origin, is only such if it recognizes the oneness of man, who is a member of the animal world, part of the living world and fraction of nature. There can be no moral code save that circumscribed within the circle drawn to include the whole human species.

Morality is inescapable because of the ineluctable dependence of species upon the environment. Without that, in the animal kingdom and in plants also for that matter, which we designate morality, there could have been no evolution. Ethics is relationship growing out of a relation that stretches through the whole to all living things, upon which is imposed the organizational history of the living world. It is an expression of the unity of nature. Inseparable from life is a structure, protoplasm, common to all living things, upon which is imposed the organization characteristic of each species. The only tangible manifestations

Page 168

of vital activities common to all forms are located at the protoplasmic surface. Whether of unicellular or of multicellular organisms, these manifestations set off each species from all others. The human morality here exposed is consistent with these irrefragable facts.

Every species-form is a circle independent of all other species-forms by virtue of species-specific protoplasm. This proposition is not debatable. All changes within this organization are consequent to behavior at its surface. Even those of the nucleus, where this be present, and of its constituents, the chromosomes, and in turn of their components, the genes, if such there be, are consequent to, and derivative of, activity at the protoplasmic surface. An animal organism, large or small, of one or of many protoplasmic systems, is ever the product of a single system, its process of emergence therefrom being one of surface-reconstitution which initiates interpenetrating catenary reactions within and which guards and guides relation to the outer world to which it responds at every step of the process. Without this differentiated living surface, the protoplasmic system could not retain its integrity, set off from its milieu, could not maintain its inward integration.

These activities at the protoplasmic surface exemplify a law, the Law of Protoplasmic Limitrophy, meaning that life, in the continuity both of evolution and of the individual, expresses itself at the boundary of the protoplasmic system in contiguity with it milieu. The Law of Preotoplasmic Limitrophy limits and restricts the life-state. This and the Law of Environmental Dependence underlie all vital activity.

Page 169

To be sure, there are races of man, so-called. But man's racial characteristics have no fundamental biological significance, judged by the criterion of interfertility between members of distinct "races". The largest numbers of biologists regard man as constituting a single species; some few, among them a well-known botanist, venture the opinion that mankind is made up of distinct species. The "races" of man exhibit anatomical and physiological similarities as close as, if not closer than, those exhibited by varieties encountered among any other of an animal species. The arguments put forward, mostly by politicians, against the singleness of the human species curiously resemble those of demagogues who, though avowing themselves Christians, deny the brotherhood of man, however far to power they come thereby; for although the Christian Church itself here as, sad to say, so often in other ways, by its acts gainsaid the teachings of the Founder of Christianity, this teaching is righteous because it is biologically sound. A man who, having a wife denies her rights as a human being, violates his own, not her, personal dignity, which is the one last treasure of every living thing. That no shame attaches to a man, who has progeny from a woman of a race which he deems brutes, makes him something less than a brute. Political, social (caste) and sociological considerations aside, we view the question as strictly biological: there are no races of man, if race means species.

As a species mankind is a unit in that human beings possess similar protoplasmic organization that expresses itself visibly by means of activities. Undoubtedly, the most distinctive of

Page 170

these activities, that which so sharply separates man from other animals, is nervous—this together with skill in the use of the hand. Brain and hand alone suffice to mark man off from the remainder of the world. We relate man's mind, his soul, to his brain, and, I add, to the whole complex of prolongations of the cells of his body, the processes of nerve-cells being but one kind of these. Then, a dominant feature of man's organizations, as in all animals, is the peripheralized protoplasm; its activities make man man. Morality as a display of mind is the relation of his species-specificity. Morality is, as is its counterpart in all animals, the index, indeed the affirmation, of his being species-true.

Similarity in activities growing out of similar structure orients the human species. It makes for like feeling, comparable to the bond common to members of any species, a feeling that is heritage of man's descent. Our heart, meaning not the pump of our blood-vascular system, but all those unplumbed aspirations for decency that ceaselessly beat within us; all those strivings for the good, the beautiful, and the true; all those unabated yearnings for ideals unattained and unattainable, are but the shadows cast by the light of our corporeal being whose sun has had otherwhere its rise. Man is the end of the process of evolution in the living world, whose course may be reckoned as but a day compared to the whole existence of the universe. In the cool of the evening, before what night we know not falls, we walk in our garden seeing in these shadows all of nature unfolded. And we feel ourselves part of all that is, all that has gone before. Immolator and immolated, votive and sacrifice, we

Page 171

bind ourselves and are bound to the universal cause of being.

Arising from a single cell, the egg, man's make-up is a unit preserved as such by a network of protoplasmic threads—the seat of his mind or soul. His feeling, reason and will are expressions of the activity of these delicate tendrils. Not only as a unit, but also as a community, does he express his being within himself and toward the outside world. But his relationship to fellow man, unlike the relations within himself involving the various bodily regions, are not due to organic connections. It is alone the similarity of structural and functional organization that puts him _en rapport_ with other human kind.

The all-pervading complex of intra-cellular connections, similar in every human being, by setting off the human species from all others, constitutes the bond between man and man. Man-to-man relationship, therefore, derives not only from common dependence upon the outside world, but also from a community of structures and functions.

Let it be clearly understood that in putting in the foreground the simple truth that all adult multicellular organisms, including man, are communities of cells, I do not intend to convey the impression that I leap to the conclusion that, as within such an organism communion is maintained by material inter-connections, so man's relation to fellow man flows along some such tracks. Not wishing to be misinterpreted, I may again call attention to what I said _apropos_ Shimer's suggestion that the gregarious behavior (herd-instinct) may have arisen from the coming together of single cells "to form many celled larger units with their enlarged possibilities".

Page 172

The intervening space between the tangible manifestations of a visible structural complex and a spiritual behavior cannot be thus leapt over.

We cannot jump from the biological fact of cellular interdependence to a sociological formula for an animal or for human society. Human sociology is in another region than biology. When we move from one region to another, from the physical sciences to biology, from biology to the social (spiritual) sciences, we keep in mind the fact that despite a common thread extending through all regions, each has a residuum peculiar to it that makes it the region that it is. All discussion of the relative values of the sciences as regions of knowledge is in this sense specious; as Compton has so well said, each level has its own value.

Human society does not create a new structural pattern as does the human egg; it manifests _en_ _bloc_ the individual activities of its members without that physical communion of cell and cell in each member. Human social communities are not webs of organic interconnections. Cellular interdependence is in one region, social solidarity in another, although no the less real, region. Nevertheless, a sociology may be formulated in consonance with the principles here exposed. The one real and true human solidarity has its source in these principles, for the oneness of man is alone fundamental. The fact that man-made social discriminations that various cults, so-called races and nationalities, which today are no more mutually exclusive, persist, does not invalidate our thesis, however much from the intensity of life

Page 173

in these molds civilization advances—for a time.

For one Milton endowed with the power of glorious speech, there are hundreds who are mute, inglorious; for every Plato, there are thousands whose lips were never worthy of the hum of bees; for a single Pasteur whose prepared mind chance favored, there are hosts by circumstances thrust from the portals where enter mankind's greatest benefactors. But in each of us the same flame flutters and in its light we move in an ever widening circle, as we must. For justice, liberty, equity and fraternity cradle our very most primal being. Do we choose to respond to duty's calls? No, a thousand times, no! Duty sits in us, created us. We are but puppets of our organization. We have no freedom from it.

The problem now posed is the source in man of his ethical behavior.

Certain expressions, moral sense, moral or ethical feeling, moral sentiment, etc., so often employed by writers on morality, indicate their position, namely, that ethical behavior derives from feeling. So clearly maintain, for example, Shaftesbury, Hutcheson and Mackintosh, to name a few. Hume also is rather on the side of theses. On the other hand, there is Locke, an antagonist of the Cartesian concept of innate ideas, although strongly postulating that ideas are products of observation and of experience, i.e., have their source in sensation, predicates morality upon law; d'Holbach and many another a French thinker derive morality from reason; and so too the greatest of all German philosophers, Kant. Mackintosh also puts the will into play. As the foregoing shows, I place myself with those who look upon feeling as the source of man's ethical behavior. The main points in my

Page 174

argument supporting this stand may be briefly recapitulated.

The capital reason for defense of the proposition that the source of morality is the feelings, is implicit in the foregoing chapters and explicit in the sketch presented on the pages immediately before this. The prototype of man's morality is that which in the animal kingdom is its _raison d'être_; animal life arose and has evolved and species have become such because of the archetype of that which in man we denominate morality. The living thing, marvel of persisting integrity, is nevertheless an exhibition of contingency—it descends _from_ the environment and preserves its species-structure in consequence of its limitrophic behavior. Its spatial and temporal projections, as state, are subject to the interminable ebullition in its medium and to the evanescent allocative behavior at its boundary. Seen thus, the state is in reality a series of rapidly succeeding events—it is eventual.

Now, at the basis of all life-reactions of animals in response to the outside world is sentiment or sensibility, as defined. This is common to all animals from sponges to man. Therefore, in the course of evolution, sentiment or sensibility appeared before not only nerve systems, organs, and tissues, but also before nerve cells. Also, it is present in any egg of even the highest animals long before the laying-down of nerve cells. In other words, nervous manifestations do not give rise to sensibility but from sensibility have evolved nervous manifestations. The archetype of man's morality in other animals is an exponent of sensibility. To repeat, avoiding circumlocution, we say that morality is the basic relation exhibited by living things. Its

Page 175

seat is primitive and enduring feeling. It arises at a lower level than reason or will, and of instinct even, for it antedates these in the course of the development of the animal kingdom or in the development of any individual animal that possess these faculties. It lies deeper than sensation, which itself has earlier genesis in animal kingdom and in man's development than reason, just as reason adumbrates will. Our ideas are but the end-points of experience, that of ours and of our ancestors, as the naturalistic psychologist, the Abbé de Condillac, held and before him, Locke. Intelligence is the fruit of sensations. Sensations are at the very base of human nature, as Helvetius said. Knowledge is refined product of sensations. In the words of La Mettrie, our soul receives everything from feeling and sensations. Morality, said Bayle, is a fundamental fact of nature. As such, it emanates from the one common property of the life-state, sensibility, or what in man is feeling. Now we are able, on the basis of our modern knowledge and on what has been exposed in this book, to see the scientific, biological basis for the statements of these thinkers, statements made from intuition or as fruit of other trains of thought.

But morality as fundamental to life implies action since action is the present indicator of its mood. Man's highly developed feeling distinguishes him from ancestral forms; its very excellence creates the need for action. Science and art, as actions peculiar to man, spring from the acuity of feeling; it is their aboriginal source. To say that feeling engenders morality is but to locate the initial impulse of action. Hence the will enters into play in man's ethical behavior. Here I

Page 176

stand with Mackintosh. But I go farther: I give a place in moral activity to reason denied by him.

Whatever the form of human action, apart from reflex-action, intelligence enters. Animal intelligence is at its peak in man. Just as the mind translates impressions into action, it also regularizes them by reason, by the intellect. Man's morality is refined by reason. To gainsay this is to hypothecate the whole splendid moral heritage given man from the past. By his greater soul man stands on the threshold of yet farther evolution. To falter is to be untrue to himself, and to the whole animal world.

Page 177

Chapter 9 - **HAPPINESS** – pp. 177-210

Happiness is the goal of ethics, say most moralists. Even a most superficial acquaintance with the history of ethics shows that the majority of writers follow Epicurus, who so clearly fixed the goal of man's moral behavior in happiness. True, their interpretations and the labels of them vary, but they spring from the Epicurean concept of morality.

To many, Epicurism means good living, a <u>bon</u> <u>vivant</u> being often spoken of as an epicure, which mostly means a <u>gourmet</u> or <u>gourmand</u> in the sense of Brillat-Savarin, who says that no language has a word that just exactly defines the man (women says Savarin are scarcely ever gourmands) who knows and enjoys good food. Although this popular meaning ascribed to Epicurism too narrowly restricts his philosophy, Epicurus himself, in saying, "The origin and root of all good is the pleasure of the belly," allows this interpretation. Indeed, he goes on to say that "wise and beautiful things are connected with this pleasure." Happiness thus is compounded of pleasures, the initial of which arises in digestion. One need not be a biologist to be able to appreciate the indispensability of the physiology of digestion, a forerunner of the process of nutrition basic to every life. An organism, like an army, marches on its belly in the sense that nutrition is basic to all vital activity. But need for food in the maintenance of all bodily functions and for the projection of the soul's most lofty aspirations is one thing, the pleasure arising from the enzyme-activity in digestion and from the proper elimination of the undigested and undigestible is another. Pleasure is not

Page 178

always a concomitant with the satisfaction of a need. To the contrary: Often, for one suffering from a gastric ulcer or from a more serious and painful malady of the gastro-intestinal tract digestion aggravates pain.

This is not to deny "the pleasure of the belly". "Good digestion waits on appetite" more pleasurable than, and quite distinct from, hunger. If you have ever recorded the hunger-contractions of your own stomach during days without food, you know how painful hunger can be. Twenty-five years ago, those of us who, as students of the well known physiologist Anton J. Carlson, made such experiments on ourselves know what pain the violent contractions of the stomach give, what profound bodily and mental disturbances hunger induces—depending, of course, on the individual's bodily and mental make-up. None of us were ever able to equal our professor's endurance, or be as keen and as vigorous of mind as of body as he was.

Appetite and the "pleasure of the table" are extremely variable and depend upon many factors, which are difficult to know exactly. So too the other pleasures, bodily and spiritual, arising therefrom. There are undoubtedly those who entertain high thoughts after plainest eating. A "New England boiled dinner" or a Swiss <u>Berner</u> <u>Platte</u>, which inspires some to noble thinking, however, induces horror in others. It has been said that with full stomachs men have given themselves over to benign contemplation of the soul; but the ascetic makes the same claim. For many men, feeding is pleasure enough: others demand accessories to the merely animal act: chosen vines of the best vintage served at just the right temperature, spotless linen and shining napery, deft

Page 179

service and, most of all, congenial company. For such as these, unlike Mr. Hardcastle in <u>She Stoops to Conquer</u>, a belly-full in the kitchen does not suffice. The pleasure of taking in food therefore varies with different peoples, with localities. It varies also in time in the history of the same people.

With respect to the amount of food taken, the mode has greatly changed in nations of western civilization during the last century. The pleasure of eating is not the same. A comparison of the menu served Eduard VII when, as Prince of Wales, he visited Paris, is in striking contrast to that served George VI last year in this same world's capital of good food—a contrast as sharp as that of the bodily build of the men of the former and of the latter epochs. A comparison of the then Prince of Wales in company with other <u>bon vivants</u> at Longchamps-night with a similar picture of last year tells the same story. In the second, one looks in vain for the embonpoint so prevalent in the first. True, there are still some politicians so ponderous that they seem grafted with fat—from over-eating only, of course!—but a pot-bellied statesman or diplomat is nowadays an anachronism. Probably no literary artist of today would subscribe to Dr. Samuel Johnson's definition of the real drink for a real man. Anyhow, the late George Saintsbury's felicitous prose in no wise suffered from his being a connoisseur of vines. Even in the wide open spaces where men are men, the picture has changed, despite one sad effect of the pseudo-arid era of prohibition, which is that the present generation cannot distinguish between good wines—to say nothing of Napoleon's favorite red Burgundy or that from the vineyard before which, in passing it

Page 180

is said, soldiers by imperial degree halted and gave the salute—and hair tonic containing fifty percent alcohol. As did their fathers of the Volstead era, the Children have tasted of synthetic gin, and their tastes no longer have an edge.

Few can deny the pleasure of eating, but fewer yet will agree concerning the ingredients of this pleasure. Paradise to some is one big "fish-fry"; to others, a stream of milk and honey. I should be unhappy in either. Each of us has his own conception of pleasure in the taking of food. And some have no pleasure in it at all. There are men who are the acme of precision and of regularity whose habits never vary; they are throughout consistent: the same food always punctually at the same minute every day, eating being a regularized routine that stamps all their activities, all their thinking. Seeing these men, I have wondered if the late Senator William E. Borah was right in thinking that men of set purpose and simple devotion to an idea have set rules concerning their meals; he must have been very irregular and haphazard where food was concerned. Borah was certainly never consistent in his political ideas. He was the Casanova of American politics—he embraced many and diverse politics and almost never two at the same time, but always, I think, sincerely.

This is an agreeable thought to read into Epicurus' idea of the goal of morality. How the history of the world would be changed were it true that peoples who have the most pleasure in food are the most moral! Chefs of the cuisine would then become more valuable than priests and professors—I mean would more generally than now be acknowledged as such. Cook-books like Escoffier's would replace cant and drivel. There would be a real league of nations not given over to phantom talk in the presence of <u>faits accomplis</u> but an assembly of serious men and women earnestly striving in healthy competition to justify each for his or her country the claims put forward by Antole France, that the French cuisine is the best in the world and its glory will shine with greater light when mankind, wiser, will put the grill above the sword. For each country has a proud right to boast of some great culinary triumph. Brillat-Savarin rates the introduction into Europe of the turkey as one of the great triumphs consequent to the discovery of America. If only he had said the same of a genuine Smithfield ham—the real thing I mean, not the imitation—properly cured and ripened and cooked with the tender solicitude, a distinction that this viand, worthy of an Olympic feast, so rightly deserves. I place it above even my own special recipe for roast capon with roasted almonds or my hors d'oeuvre of frog eggs—<u>eggs</u>, not <u>legs</u>. Cooked with apple cider of its native Virginia, it is for kings, nay, again according to Savarin, even for Cardinals. For special though Burgundy is, the red "married to Roquefort" is classic, excellent though it be for the ham of Prague, alas, now no more, the forthright flavor of Virginia's pride demands the juice of the apple. You who know the succulence of this morsel, have you ever stood before a shop-window not far from Piccadilly Circus and seen, among the hams of this earth, a Smithfield holding its rightful place in the very center? And have you not been carried away on a wave of nostalgia, moved to exclaim "this is my own, my native land?" Sole Normandy style in Rouen, foie gras in Strasbourg,

Page 182

partridges in the hunting season in Colmar—these have truly righteous claims. But the ham of Old Virginia is America's vote in the assembly of nations.

A legendary Harvard professor, no doubt of well-advanced age, has affirmed his faith in the superlative value of what were he a physiologist he would doubtless have christened, as earlier two famous English physiologists did christen, the Law of the Intestine, referring to the movement of food progressively among the gastro-intestinal tract. Intestinal stasis, a condition violating this law, may have dire consequences. You doubtless recall that one of Voltaire's characters, an Englishman, discoursing wisely and seriously, as his characters always do, proves that bad digestion in a ruler may determine the fate of a nation. It is said that the first Napoleon was a very rapid eater. He paid for this in the end, his star falling as his stomach failed—at least this suggestion has been made and is perhaps as good as any other. Napoleon, dying, exclaimed his gastric distress without apologizing for taking an unconscionable time in the process. Thus lack of the pleasure of digestion and incomplete digestion may be frought with serious consequences.

But all this notwithstanding, however much we give place to taste and to smell attending the "pleasure of the belly", we are forced, be it reluctantly, to discriminate between the pleasure of eating and the need for food, between appetite and hunger, and between enjoyment of a process and the satisfaction of a demand. Philosophy has yet to decide the question of value, which seems suddenly to have become, according to several philosophers, its central problem. Less can psychology or biology

Page 183

estimate the relative value of pleasures. Ordinary methods of measurements, those of intensity and duration, fail. Our ideas of pleasure are perhaps as individual as fingerprints. Not so biological, that is, vital, needs. These are common to all mankind. If we propose a morality whose roots are in biology, in the very organization of the life-state, this morality has to be consonant with established biological principles, those for which there can be no dissidence. Hence a morality originating in the fundamental and inescapable process of nutrition is ectopic; one predicated on the pleasure experienced in the process, which may or may not be present and if present varies so widely, cannot apply to all mankind. Most certainly I do not gainsay the pleasure; I do deny that it is common to mankind in being everywhere and having always been the same. It would seem to be scarcely worthwhile in the light of common knowledge to dwell further on the fact of the imperative need for food or on the consequences of intestinal disorders.

You may here justly say that the Epicurean morality of happiness does not rest on the pleasure of digestion alone. Happiness has here its beginning. According to Epicurus, all the pleasures taken together comprise happiness. Moreover, happiness connotes the absence of evil and the concord of self with others. Each human being who is striving for the same goal will by mutual agreement avoid inflicting suffering or harm—and so avoid receiving these. Further, it may be objected that the Epicurean idea of morality, which has been so extensively developed since the time of Epicurus, ought be widened to include modern conceptions other than a restricted hedonism or eudemonism. Epicurean morality ought to include

Page 184

the aspiring to happiness by way of pleasures to body and soul.

Pleasure of eating often must be subordinated to some other physical enjoyment, for instance, athletic sport. An athlete in training puts himself under dietary restrictions, limiting his desire for favorite dishes and concoctions which, though savory, lack brain-building power. Yet he is well-fed, no doubt of that. How most of us, poor anemic ones, "greasy grinds", as the football heroes of our college days, the only real heroes of academic life, contemptuously named us, gazed with sad longing upon the envied platters heavily laden with tender specially selected roast sirloins of beef that passed through our large dining-hall to the smaller rooms where these gods sat at "training-table" whilst we dined, true to character, on corned beef of uncertain age but not uncertain brawn; or upon the jugs of thick rich cream accompanying the bountiful breakfast-trays, as we swallowed tepid tenuous coffee tinted with "blue milk" and attacked soggy doughnuts stinking of the ancient mutton-tallow in which they had been fired. Albeit vicariously, the non-football player enjoyed the autumnal hectacomboid sport, subordinating his pleasure of food by being party to real glory. And all the while forgotten, those early days of each beginning college-year glowed with just a hint of winter yet to come, to weave a white and lacy mantle in the ruined choirs of the trees now brilliant in red-golds and browns spun from the shuttle of autumn—nature's loom! How it adorns each season's mass.

Page 185

Football is no extreme example. Athletic prowess demands training and this means subjugating the pleasure in eating to some extent. According to Bentham's "arithmetic of pleasures," this is correct; one pleasure is sacrificed for the sake of another. We are often virtuous in some ways for the sake of greater power for indulgence in another. Socrates once said something like this. But this pleasure in physical prowess varies almost indefinitely. The amateur athlete, he that is honestly not paid for his services, has one standard; the quasi or semi-professional, the occasionally indirectly paid, has another; and the true professional has yet again a different. Heyward Broun, I think it was, once said that the professional athlete deserves as much esteem as the amateur. Speaking of artist, but using the terms amateur and professional in another sense, Guyau tells us that the former is egoistic, the latter not. This may apply to athletes and to sportsmen generally, I think. Take fishing or hunting, for example, or mountain climbing. The lust to kill is certainly quite different from taking game for need; the fanatical mountain climber who risks for the sake of risks is in another category than that of the paid guide who protects and rescues lives.

Many men hardy and vigorous enough to indulge in the grueling contest of rowing, in basketball with its shocks from quick movements and halts, or in the god-like sport of prize-fighting—for as a friend of mine once said, Hercules, the first of the world's great champions at the heavy-weight, was made a half-god, as they are now—prefer the milder pleasures of fishing or mountain climbing though they catch no fish or scale no difficult peaks. The enjoyment of

Page 186

a sport is thus strongly individual. Even when physical strength and endurance more than suffice for the most arduous games, a man may prefer the less strenuous sports.

Crossing a glacier under the leadership of a guide in a company of about twenty persons, speaking Italian, French, German and English, I was struck with the difference between the pleasure of such an excursion as the means of healthy out-door exercise under burning sun on the thickly packed ice among such polyglot society, and the pleasure that comes to those who risk their lives in the sport of climbing dangerous icy peaks. At the halfway point, as we paused for ten minutes of rest, the guide looked through his glasses onto the highest pinnacle. Each of us in turn looking could distinguish three human forms madly dashing first in one direction and then in another as they sought escaped from an island of ice. Around it now was a gap too wide to leap. These people had unwisely climbed up on the sunny side of the glacier instead of its shady side, thus leaving out of account the effect of the sun in making ravines. I was never able to learn the fate of these people. The guide could only say at the time that rescue was impossible.

Guyau makes a strong point of the pleasure that comes from taking risks in games, citing animals that play with death in their sports. However, a man may not be a coward who prefers to risk life and limb in what to him is a more serious undertaking than a game. Moreover, the value of enterprises that engage us changes. No man is perpetually a second-year college student. With the years, what were once values cease to be—even for

Page 187

the "inspirational" college professor who is one of the boys; he is often a case of arrested development, one who never knows what Guyau names the metaphysical risk.

Like the poppy-bloom, pleasures often vanish in the taking. It is just as well. Man can never withstand the continuous impact of pleasurable sensations. Continuous emotional display is impossible. One of the highest human pleasures, that of friendship, based as it is on mutual understanding and answering trust, as Henry Churchill King says in his thoughtful book, <u>Reconstruction in Theology</u>, must not expect continual emotion. So too in spiritual enjoyment, where again the criteria are so various.

What are usually called the physical pleasures, meaning the bodily ones, are difficult to reduce to a general formula. Individuality, a complex here of physical prowess, agility, keenness of vision, and nice response of muscle to nerve-impulse ("timing" or, in physiological expression, reaction-time) which plays so large a part in games where a ball is used and in boxing and of personal inclination—plays so preponderating a rôle that generalization is out of the question. Take the many games played with a ball—baseball, football, tennis, golf, etc.—in these, mere physical brute-strength is not enough. Agility, keenness of vision and especially what in the parlance of sport is called "timing", and in physiology reaction-time, the rapidity with which muscles respond to the nerve-impulse, count. An agile player may be more valuable than a more powerful one. Few athletes lack keen eyesight: seldom does one meet with a spectacled baseball player. Keep your eye on the ball is a fundamental rule in all games in which a ball is used. "Timing" in tennis as in batting a base-ball or receiving the foot-

Page 188

ball is essential, and the player who most excites our admiration is most often the one with an almost unbelievably short reaction time. A pugilist who "telegraphs" his blows is lost.

Pleasure in the use of the body is thus contingent upon reactivity, which, though a human propensity, is extremely variable. And there is always the predominant factor of personal inclination. Moreover, many a skillful athlete gets no pleasure out of games at which he excels.

Pleasures of the mind and spirit are no less varied. One man's supreme enjoyment is in scientific work; another's in art. Others derive pleasure from appreciation rather than from creation. Scientists, even in the same field, differ with respect to the pleasure that they gain in work; some few even deny that their work is pleasurable. The production of a thing of art is often painful to the artist who paints a picture, composes a sonnet or a symphony, or releases from a block of marble an imperishable dream of beauty. I know a man who could have been a champion rugby-player, judged by his success in casual play with veterans of this sport: that his greatest pleasure is in higher mathematics, his erstwhile companions in play could scarcely comprehend.

The pleasure in creation is apart from praise or blame, money or glory, but rather lies in the joy of doing and in the self-generating good feeling of self-expression. Creation therefore is its own reward, made richer, of course, by the pleasure it gives others. But the creative artist or scientist never has his eye on the chance of approbation; his eye is single and fixed on the bringing into light his dream or interpretation, failing the accomplishment of which he suffers the torture of the damned. Creation

Page 189

is thus a god-like talent by chance implanted in women and in men outside of any law that we know how to formulate. It is there or it is not. It may or may not kindle in others the spark of appreciation; when it does, it is universal and leaves the sphere of special creation to become the embodiment of all mankind's yearning. The highest form of creation is self-abnegation, the sacrifice of self to a divine frenzy to get rid of a spiritual burden. It is thus less a pleasure than a surcease of suffering. It is a self-cashing out of the devil.

Needless to say, appreciation is more widespread than creation. To few is it given the ability to elevate the symbol of truth and of beauty to a higher place; many may have their souls uplifted in fixing their souls onto the symbol—the higher it is, the more whom may view it. What counts is that, upraised, the symbol now becomes common property. We fix our eyes on the thing, not on the hands or shoulders that bear it.

Each of us has his own peculiar pleasure in appreciation. Banality of banalities. Without rhyme or reason, a picture, a landscape, a building, a piece of sculpture or a strain of music invokes response, never quite the same in two persons or in a single person at different times. To me, the more enduring forms of art are more appealing than the more perishable. I have often said this to myself, until I hear some music that enchants. So, the temple of Paestum, for all its suffering from the ravages of time, evokes a greater response than a painting more susceptible to injury. The victory of Samothrace has greater grip than Mona Lisa. Yet I feel now that such cataloging is false. The time-evading quality of a work of art is an illusionary criterion.

Page 190

Each form has its own limitations fixed by its medium; architecture or sculpture, painting, poetry or prose and music are to be esteemed and appreciated each within its sphere. Again, seeking repose in the appreciation of art, I had thought before I saw it the Laocoön group would be horrible in its restlessness, and that the sculptor ought not to endeavor to portray the happening of a moment, especially one of extreme agony. Seeing it for the first time, feeling even greater revulsion than I had expected to feel, I was persuaded that momentary and intense movement cannot be portrayed in sculpture. But there is the Niké Samothrace poised at the top of a stairway in the Louvre. How at the first sight it moves one almost to tears, its outstretched wings seem to beat the air, carrying one back through the years and making one part of some infinite glory of being. It is the poetry of Homer, in all its simple imagery and rush of pleasant sounding words, as the Laocoön is the Aeneid, forced and harsh.

The spiritual pleasures, in aesthetics, science, morality and religion, vary as much as the physical. In art, it is a difference between creation and appreciation, between the modes of participation, between the idea in the artist which he expresses and the beholder's interpretation of the creative idea and embodiment of this idea, and between the kinds of activity engaged in by creator and appreciator. These differences vary enormously with each form of art—the literary arts as well as dramatics, painting, tapestry-weaving, sculpture, architecture and music. But in all forms of art the appreciator partakes directly of the thing itself; what he sees and hears is the thing itself. With science it is otherwise.

Page 191

A scientific achievement becomes quickly common property, and all mankind may enjoy its benefits, the more so the more widely the knowledge is disseminated. But the act itself is reserved to the discoverer alone. There is no partaking here, albeit there is acquisition. A scientific act of discovery can never be shared with anyone.

Art is tentative, hesitant even, and, of all spiritual expressions, the freest of dogmatism. And it is this naiveté and uncertainty that are its charming appeal. Its very medium restrains it, and gives it a sort of helplessness that is direct and heart-warming. The artist may dogmatize, but his creation never does; the critic may analyze according to his theories, yet the thing criticized remains refractory to subjugation—even a mosaic, most of all a mosaic. What offends in the cinema and hinders it from becoming great art is its lack of restraint; it is a medium without confines of time and space—the very antipode of the unity of the Greek drama. The "cut-backs" and the "close-ups" and the superposition of scenes to reveal the memory of the actors, aside from the moral value of the cinema as widespread purveyor of bad taste, indicate how much art depends upon limitations for its appeal. If dramatic art such as a play is naively childish, the moving picture is childishly sophisticated in a rowdy fashion. True, each form of art has its rules, whether it be poetry and prose, sculpture and architecture, painting and tapestry making, or music and the dance. But the most fundamental rule is that of inescapable allegiance to the medium of portrayal. This, obviously true of music, a time-thing, is just as true of a stage-play that can never leave the scene of action.
unless restrained by the taste and talent of the producer. Chaplin's Woman of Paris remains the best film that I know. Chap-

Page 192

Music captivates by its sheer impermanence; architecture and sculpture by their time-defying quality. The Magic Flute is spiritual exaltation because its music is temporally woven magic being free of space, whereas the temple of Paestum and the Aphrodite of Melos, for all their losses, give us the feeling of persistence in space and time. Music stirs poignantly or beatifically to exteriorize our submerged best selves; sculpture and architecture by the very quietude of their medium, create a restfulness whence emerge strivings for activity in stricter accord with beauty. True, sculpture is often a portrayal of the almost immeasurable action of the moment—but often it fails thus: the Laocoön group, for example, which falls short not only because of the frozen horror of a moment held captive in stone. Yet, momentary action has been thus imprisoned to our delight. There is the Nike of Samothrace, poised on the landing of a stairway in the Louvre, whose wings seem to beat the air. The rankest skeptic risks to believe in angels in beholding it.

Thus, art is all-embracing. As a living thing, more than a living thing, it transcends time and space. The delicate tracery of an aria hands for a moment in time, unsubstantial in space, and is gone, whereas architecture and sculpture, time defying, persist only in space. Music is like incense, a divine vapor rising from the soul to create soul, the breath of eternal creation. Sculpture is the permanent realization of dreams.

Page 193

What egoism is, it is difficult to say. Specific instance which, it has been suggested, are sometimes the best of definitions, fails, inasmuch as no single example drawn from writings on egoistic morality suffices. Attempts to reconcile them are fruitless; superpositions of them confuse. Putting together the various ideas expressed by moralists on egoism, one has not a picture in proper focus but a mosaic of contradictions and oppositions. There is Helvetius with his law of self-interest against God-drunk Spinoza, the gentle even humble egoism of that strange mixture of skepticism and conservatism. Montaigne was so unlike Rousseau although both believed more in man's nature than in education. But how unlike again their estimate of this nature! Think yet of the English utilitarian school which traces back to Aristotle, of Jeremy Bentham with his scales for pleasures and for pains, and his arithmetic of happiness, and John Stuart Mill, and of their predecessor, Shaftesbury, to whose school Darwin belongs. How typical of the hardheaded nation is Mill's ethics, at once the product of the logician and economist and of a people who, so to say, invented economics to be an accessory to their national genius, politics of practicality, of realism. Hippolyte Taine, great admirer of Stuart Mill—we are told by Maurois that Gallieni, chef of Lyautey of Morocco, when they were in Indo-China, loved to read Mill's Autobiography—seems to have had right, in applying his trilogy, of race, milieu and epoch, set forth first in his <u>La Fontaine</u>, in his <u>History of English Literature</u>, that ideas are determined by material conditions.

If we attempt to comprehend egoistic morality by sorting the various writers, and arranging them into categories, lack of

Page 194

success attends the effort. Thus, there is no common ground to be discovered whereon the naturalistic moralists, in whom we have special interest, stand. Rabelais, man of medicine and teacher of anatomy at Lyon, where he made demonstrations on the cadaver, curé of Meudon and diplomat, the antithesis of Calvin who outlived him by eleven years, as moralist is follower of Epicurus, a wise, optimistic and jovial naturalist. To appreciate him, one has to go below the surface of his scabrous writing to which La Bruyère voices objections. He is against enemies of nature, those who falsify it or put it under constraint either by vice or by narrowness of spirit, against pedants and the people of the Sorbonne, against men of the law who briment and tondent, against monks fainéant and living in luxury, and against brettenas and men of war. He loves knowledge and liberty.

Pierre de la Hamée, called Ramus, contemporary of Montaigne and of Charron, and victim of St. Bartholomew's night, was an anticipator of Descartes' deduction method. Holding that reason ought to remain at the school of nature, he nevertheless as moralist made certain reservations: one ought not to abandon one's self to nature often bare and vulgar.

There you have it, and would still were you to extend the list.

Similarly, following the lines of influence, we do not come to a junction of ethical theories. We cannot say of the history of morality that ideas streaming from various sources unite at last in a common bed of thought, as it has been said of the egoistic morality that "the virtues lose themselves in interest as rivers in the sea". Turgot, himself economist and minister of state, believing in the land rather than in commerce as the source of a country's richness, influences Mill: to Shaftesbury, Hutcheson and Rousseau, Kant acknowledges a debt; yet, little have their philosophies in common. Ideas gather force as they move, <u>are</u> force, but are transmuted in time and in personalities; one may become a glorious ideal of soulful endeavor, the other a mere ideology which, through an individual of atavistic make-up that shames the brute-world as procurer, becomes spiritual proxenetism. So is the idea of egoism, both virtue and vice.

The moralists' conception of egoism failing to satisfy, we turn to biology. What biology has to say is another matter. At once it rejects egoism as vice. Life as revealed to us, as we have seen, rests upon the sustained efforts of definite structures to maintain themselves in their surroundings. This sustentation is incompatible with vice. Life, by its overwhelming capacity to persist in conformity to the laws that dictate existence, is virtuous, cannot be evil. Egoism, as an expression of individuality, is the inalienable right of amoeba and of man, the right to thrust itself forward in the milieu, to maintain self, to survive. But biology also teaches that individuality is a puzzle.

Individuality, the highest expression of uniqueness, the <u>infima</u> species of specificity, in the world of living things is as unresolvable as are chemical elements which, although when broken down show themselves each a combination of electrons, are nonetheless individual since electrons are incapable of independent existence. We measure and weigh the electron, estimate it in various ways, including the estimate of its quantum of energy, jumping from it, so to say, much as does a smoke ring, which condenses at a certain distance from the orifice where it

Page 196

issues, non-existent as a ring until it forms, whose locus of formation may be more approximate to the point of issue, as with steam. Thus, the quantum of modern physics may not be a reality but an expression of the activity of the electron in changing its energy with, or consequent to, the change. The single living thing is the most individual in existence, for though each of any species is born of the same combinations and obeys the same laws that maintain the species, each is a thing peculiar to itself. Two examples of a chemical made up of molecules or atoms are the same, must be the same. Two living things never are. With respect to animals best known to us, these statements are banal.

Every practicing physician knows that human beings differ widely in tolerance to drugs, and in susceptibility and resistance to disease. It is almost true to say that one man's drink is another man's poison—true of milk as of alcohol. Scarlet fever or other contagious disease may sweep through a village or a family even attacking some, leaving others untouched. Whether age, sex, weight and size, body-surface area, nervous condition, glandular efficiency, or what not, having duly taken these into account, there nonetheless remains somewhere in every patient some idiosyncrasy arising, if not from some peculiarity, then from the individuality of the patient as such. Medical practice is in large measure an art and will remain so until we uncover the mystery of individuality. This means no disparagement to medical science—far from it. The advancement in medical research is one of the great achievements of modern science and, though made at the expense of professional biology, which so often follows now instead of leading experimental medicine, any biologist is ready to acknowledge medical triumph. Nevertheless, the human

Page 197

patient is never a guinea pig, or a white rat or a fruit-fly, however excellently standardized "laboratory material" these creatures may be. The step from the clinical laboratory, to say nothing of that from the animal hutch, to the bedside is a long one, to be taken by a trained man, yes, but also by a humanitarian who reckons his patient as something more than a subject to fit his card-indexed notes of scientific learning. This is why I believe that state medicine or social medicine or the "group-medicine" of high-priced specialists can never replace the old-fashioned family physician. Social medicine? The very term hypothecates the physician's duty—to treat every patient as human individual, a complex not only of bodily ills, but also of spiritual substances.

Human individuality being superimposed upon species-specificity, human diseased conditions of course have the same substrate, the protoplasmic organization. It is not too much to prophecy that we shall yet master the individual differences encountered among men in their response to drugs, and germs, and understand more thoroughly and basically the pre-disposing causes to disease. But first the larger biological problem, of relating species-specificity's inherent indissociably with the intimate structure of protoplasm, will need to be solved. Signs are not wanting to indicate the solution of this problem as I have elsewhere pointed out. This problem answered, it will not be difficult to attack individuality as manifested to the practicing physician. Once this question is posed, one thinks immediately of feasible methods that could lead to its answer.

Despite the normal variations in bodily structure—there are

Page 198

others besides the useful fingerprints, the course of the superficial veins, for example—and in bodily functions there is a medicine obviously that applies to all mankind. Medicine as biology must reckon first of all with the fact of species. From this point of view, medical science as such takes little account of the individual revelation of nature's seemingly endless capacity for producing variations. But the individual rests forever, as far as we know today, within the orbit of the species since no variation crosses the species-frontier. For living nature, then, the individual means little; what counts is the species. Biologists stress individuality less than the species because of this rather than from ignorance of what individuality and individuation really mean. "Rugged individualists," of which the political barnstormer s love so much to talk of, is a "side-chain" of species-specificity. The universe as a whole, living world as a whole, animals as a whole, and species as a whole—those are primarily the units of science.

If one could estimate in worth according to weight, measure and number, if there were a quantitative determination of good, chance, and Bonheur, then somehow the abstractions of physical science would still counterpoise the persistent human cry to be happy. But science, physical or otherwise, gives no hope that we may yet put spiritual longings on scales, alongside a yardstick, or reckon it with an adding-machine. Science can and does alleviate sufferings and mitigate hardships, and these it does without having promised anything. To reproach science that it has failed to bring the world Bonheur is also to reproach art and religion. Quantitative science does not counterpoise the aspirations of the soul.

Page 199

The equation, pleasure plus individual, i.e., egoism, = happiness, containing two unknowns, pleasure and egoism, may or may not be correct, and is indeed correct only if another factor enters. This factor is chance. Happiness is a chance phenomenon depending upon the named pleasure and the individual. Hence, our equation is true or false depending upon a variable. There is nothing either in a pleasure or in an individual which, when brought together, add up to happiness. The digits in an arithmetic of pleasure have so inconstant a value that the arithmetic is useless; the addition of the $\underline{x}$ of personality, the egoistic unknown, only increases the difficulty because this $\underline{x}$ represents in each case one individual of the entire human species. There are no stoichometrical laws for happiness: a fixed amount of pleasure plus an established egoism equal happiness. Rather, the variables, pleasure and individual, give a varying compound, happiness.

Without insisting too much upon the meaning of words, we nevertheless need to know, for the purpose of clarity, how in any discussion one defines the word, happiness. Inasmuch as words have often strict as well as one or more looser meanings, we endeavor to eliminate connotations and implications that may color a term. Etymology, when known, though it does not always serve our purpose, since derived meanings implanted by usage may replace the original, the English language, as Greenough and Kittredge have admirably shown, abounds with words with many, and therefore hazy, meanings. Whenever I use the word, happiness, I recall my boyish lessons in rhetoric, when I used Genung's textbook [Genung 1900]. There a distinction is made between happiness and joy. Happiness, a sensation that comes from a happening or an event, is to me something quite different than joy.

Page 200

A French equivalent of happiness is <u>Bonheur</u>, good chance, from <u>bon</u>, good, and <u>heur</u>, chance, although <u>Bonheur</u> means perhaps something else than <u>bonne chance</u>. The flavor or color latent in the word, happiness, is for me chance, luck. If for you the word has the same connotation, we are in accord. But you may disagree, in which case you do will not accept this argument. Then we rest on the earlier point, namely, that despite this (to you) invalid distinction between happiness and joy, happiness is a fortune determined by the incertain reactants, pleasure and the individual. Quite apart from dictionary-definitions, I cannot escape the feeling that happiness is the result of chance.

Does chance then determine the goal of ethical behavior? This would seem a barren morality, that in our actions we are creatures of hazard. I reject any theory that postulates happiness, which depends on the play of chance, as the reward for decency.

Let me hasten to add that I do not thereby accept the pessimistic morality of Schopenhauer, of von Hartmann and of Mailänder, according to each of which life is a dolorous undertaking, and that there is only suffering. But these mean apparently did not believe in the systems of morality that they set up in their youth, except Mailänder, who alone committed suicide. Without denying the suffering and misery in this world, we need not embrace them as the cause of our being. Even the "happiness-moralists" sometimes make the mistake of presenting the problem in the form of alternatives: happiness or suffering. This is false. Not to be happy does not mean to suffer. Minus pain does not add up to pleasure any more than minus pleasure equals pain. There is a middle-zone both in physical and in mental states where neither

Page 201

is present or felt. So it is with happiness and suffering. Just the same, I do not advocate the doctrine of passivity, of nirvana. The recognition of a middle ground does not demand its being taken. Morality has got to be positive and affirmative, neither neutral nor negative. Hence there can be no middle zone of temperature where life is quiescent if not moribund. Life is quick—in the sense of unceasing and rapid reactions. We are not pessimists or optimists in our morality because morality has no point of view.

According to Schopenhauer, there is only one single innate error. This is to believe that we are here to be happy. It would be more just to place the end of life in our sufferings than in our happiness. All human existence indicates suffering as its true destination. Death is the true end of life. There is only one means to resolve the contradictions of life. The eternal principle is the idea of the species or of the genus; this principle remains indefinitely (Cf. Spinoza). Nature conserves only the species. We ought to put ourselves in accord with nature and to consider life plus death as entirely indifferent. The supreme end is negative, resignation.

Eduard von Hartmann later shared Schopenhauer's opinion of the impossibility of considering happiness as the goal of life. His arguments were: 1) One may think that good luck can be acquired during a life-time—youth, health, love, glory. But these fade as a dream. Love, for example, brings as much pain as pleasure; therefore, von Hartmann is against love. 2) Immortality. Lacking good luck here, we hope to have it after death—another illusion. 3) The third illusion: progressive development of the universe. The happiest people are the most primitive and those that are the least knowing among the

Page 202

cultivated. The more we learn, the more we are discontented. This is the best of all possible words, but it is miserable nonetheless. So, to von Hartmann, Nirvana is the way of escape.

Accepting Schopenhauer's point of view of the miseries of life and von Hartmann's stages of illusion, Mailänder puts himself against Nirvana and the will to live. The world to him is a means to an end, the non-existence to which it moves. Life is a step toward death; to appreciate it, we live. Whereas Schopenhauer and von Hartmann lived on long after publishing their essays, written in early life—Schopenhauer's at the age of 31 and von Hartmann at 26—Mailänder committed suicide at about 35.

Three reasons push me against this philosophy whence derives a negative morality.

There is first of all the error in setting up suffering as the alternative of happiness, whereas the absence of one does not imply the presence of the other. Because one is not happy, it does not follow that one is suffering. The absence of and even the release from suffering does not mean happiness any more than lack of happiness brings one into the state of suffering. Thus, instead of two, there are four possibilities: happiness, no happiness, no suffering, and suffering—four instead of three because there is a difference between not happy and not suffering. The surcease from bodily or mental suffering is not the same as the state in which happiness is lacking. An anodyne relieves pain but does not of necessity give euphoria. An anesthetic may dull pain and yet give a feeling of discomfort. An injection of an anesthetic into an aching tooth to be extracted, though it deadens the nerve, by no means gives pleasure. I can imagine nothing

Page 203

worse than the absence of all sensation, of touch, for example—what would I feel in sitting, standing or lying? The obliteration of sensations, even of pain, is not of necessity pleasant. That pain is alleviated does not mean that pleasure is felt. This I would say applies to mental as to physical states.

Secondly, there is in these pessimistic philosophies the error of regarding repose and annihilation as synonymous. A body completely at rest is not non-existent. To wish repose is one thing, to wish to be nothing is another. Rest, absolute abatement of activity, were such possible, may be the very opposite of disintegration. Further, with respect to death as the weapon of nihilism, one can only venture opinion. It may or may not be such. When, moreover, Haeckel, for instance, tells us that the tomb is the only place of complete repose, he speaks no longer as a scientist, but as a demagogue of materialism; he has made no experience with life in the tomb—I mean as a dead person. He too confuses annihilation with repose.

But my greatest objection to Schopenhauer and his followers is grounded in biology. Life only figuratively is engendered by death; in our actual experience, life is produced by life—a point to which we have earlier referred. But suppose for a moment that we admit the figurative as experience, saying that because animals live upon others as food, life emerges from death. This can only mean: some certain forms of life persist because some other forms cease to live; it cannot mean that in the death of organism A, organism A lives. And that is the point involved—not the continuity of some lives at the expense of other quite different ones.

Page 204

[Box 125-21, folder 396 contains seven pages handwritten in blue
ink with, at top, pencil-circled numbers 23 to 29.]

[Circled 23]

If you say that death engenders life, you obviously mean that
the death of one organism maintains life in some other organism
using the dead for food or by killing it out of self-preservation. You
do not imply that out of dead A, A's life is engendered. Thus, the
assertion that life emerges from death applies to two different
organisms. To the extent that it applies more generally, to all
carnivores, for example, which eat herbivores and to the latter,
which eat plants, it is a law. Then in like manner, it is entailed that
the next step logically be taken, namely, it is further entailed that
generally life is maintained. Thus, we arrive at the conclusion that
death is an individual phenomenon. I mean: the insistent fact is the
persistence of life. Though countless members of animal and plant
species have become extinct, life goes on today in countless other
members. Death is the codicil, the final word, to individual life and
to individual lives; up to now it has not involved all living things.

One other point. Can we argue from the end of life unto its
beginning? Life ends in material dissolution. To say that when a man
dies his body returns to the chemical elements that compose it, is
trite,

Page 205

[Circled 24]

commonplace. But, however far back we trace this dead body to its beginning as an individual, we do not, cannot, come to a point that has anything in common with the elementary dissolution in death. Only by convention, for convenience, do we speak of a life-cycle as meaning the history of a life. Actually, there is no cycle in the sense that individual life returns to that whence it came. The disappearance of life is one phenomenon; its initial coming into being is another. Life begins for a human in the egg, but this is already alive; with birth, the life trajectory has already long ago begun.

Page 206

[Circled 25]

We are not pessimists or optimists in our morality because morality has no point of view.

There is something inexorable about morality. It is thoroughly inflexible, relentless. It is there, as is life, and we make the most or the worst of it as we do with life. Hence a pessimistic or an optimistic morality is as untrue as a pessimistic or an optimistic life. I mean: biologically speaking, life either is or is not, either moves or ends; but it, as it exists and persists, is neither an object of pessimism nor of optimism, and we err when we read into the state of being alive the jaundiced state of our mental outlook or exuberance of spiritual well-being. Morality is apart from our attitude to it, as life is apart from our view, which may vary from one moment to another. Yesterday during writing I paused to observe among a culture of fifteen-day old larval sea urchins some Amoeba flowing forward out of the field of my microscope, as if endowed with purposeful activity. Their pseudopods thrust forward seemed like bubbles filled to the bursting point with the joy of living. Today, due to various factors among them, including the great difference in temperature, these tiny particles of life are sluggish. But they were not optimists yesterday, and they are not pessimists today; the tempo of the life processes differs, that is all. Should an observer suffering from one of those states of depression that comes to us all at times read into the Amoeba the observer's mental state? I cannot see how a pessimistic or an optimistic morality can interpret that which persists quite apart

Page 207

[Circled 26 and "-6-" in pen]

from our being, be it that of a Rousseau, or a Schopenhauer, or of us all, were we one or the other or equally distributed between both.

As the amoeba, so the egg: it makes no question of aye or no, of optimism or pessimism, but straight along the path of its development it goes. And in going, it "knows" what it is about – I mean it responds to its environment moment by moment and in this wise elaborates by itself, according to its specific organization, that which becomes finally sea-urchin, sea-bass or man. In this manner of speaking amoeba or egg is moral: it plays the game as it knows it according to rules, the only game that it "knows" without feeling of pleasure or pain (presumably) without hope of reward in amoeba or egg paradise here or now. In each case, life is lived. And this living is its own reward, its own joy. Life, we have said before, is an event, or better, an unending series of events. But it never simply just "happens." Even in the coming into being of the first form of life we may assume that something more than chance dictated this play. At any rate, the persisting qualitative tangible substance permits us to affirm that, whatever chance there be assumed, there is a very legal and law-abiding deus ex machina which, because of its limitations, carries the fictitious name, chance.

218

[Circled 27 and "-7-" in pen]

What I want to say is that a morality, certainly one based in biology, that puts happiness, however defined, as its goal is neither moral nor biological. There can be no moral value in an activity induced by hope of reward. There is no biological phenomenon that catalyzes itself toward a chance-end. The unending specific re-duplication by structure is the prevailing, obtrusively prevailing, datum that biology affords. Without this persistent true-to-structure principle, it would be difficult to envisage a world of living things. But if, however, your imagination can picture such a world, good! There nevertheless remains the obtrusive datum afforded by our experience that the world of living things is composed of life-forms put into species whose members are species-true. And they are true not by some chance, remain true not for some chance, and continue true through generations outside of any reward save the living of the life itself. This "species-trueness" is for animals their morality. It _is_ man's morality.

Page 209

[Circled 28]

Happiness is a state of mind, and as such it is as much in the past as in the future and not only in the moment. It is active and indicative, but may be in the past, present or future tense. The projection of the mind into the future of activity, memory of action or thought, may be greater than the good moment seized. Suffering as remorse and as feeling for another's ills is quite different from mental or physical suffering in the moment. Therefore, suffering and happiness are not antipodes: the measure of their latent and refractory periods differs: the traces in the memory of Happiness are not comparable to the memory of suffering. The after-beat or overtone is different in each.

Page 210

[Circled 29 and "-1-" in pen]

The eternal quest for beauty and the interminable search for truth are fugues without end in the soul of man. The lyre has yet to give the last and final chord, the retort to distill the primal, universal essence. We walk as in a dream, forward and upward ever pursuing, never attaining, the ultimate goal that constantly eludes us. But the quest goes on, the search is relentless, for there is ever the hope, though delicate as a mist, an intangible vapor, that beyond, just beyond our reach, is the goal that man's heart sets him. Like dreams? Perhaps. But the infinite glory, the unattainable good, lies in these dreams, these shadows of a long past that bid us move on to meet our ensuing destiny foretold for us in the first primal ooze of this earth where came to be the first respiring life, itself the primal ooze of the living world. Mists enshroud the far-off mountain top where perhaps dwell in purity all that we would attain. Toward these heights we labor we know not why except that the way was predestined long before we came to be. An army marching in parallel lines that in some long future converge, as evolved to man the living would, we break no ranks but, true to our heritage that is our destiny, we toil on. The glory that is alone worth the hazard is that born of our being. The effort is the only worth, the right of travail, yes, and of suffering even is the gift that life gives. Living is the only goal. To live is reward enough of life.

Page 211

[Box 125-19, folder 382 contains alphabetized pages a through z, plus z 1 through z 8.]

a

Postscript - **MUTUAL AID** AND **ETHICS** – pp. 211-243

The promulgation of the theory of natural selection, emphasizing as it did the struggle for existence, dates a crisis in ethics. Over-emphasized by Darwin's followers, the struggle for existence came soon to be a credo not only in biology but also outside of it. It gave birth to a philosophy that founded a new political school of thought which, in my judgment, came to be more pernicious than the Machiavellian idea which, save for sporadic recrudescences, was in modern times outmoded. Thus translated, it gave western civilization a new fire, all the more injurious because an invention alleged to be nature's creation. It came to be spiritually a burning of all books about airy dreams wherein Utopias and sequestered isles of peace were chartered. It left only hell burning, made earth a hell of struggle; in the smoke, paradise vanished and with it peace on earth and good will to men.

As Jean-Marie Guyau has said, the dream of moralists is to find an antinomy for this law of the struggle for existence. This, as the hope of Utopias of an earthly paradise, may be in vain. And yet we yearn to see something positive in place of shattered hopes; hence our dreams, faint shadows of hope infrangible. At least if paradise must be put to flames, we ask that hell be inundated.

Curious fact of history, this, that the Darwinian theory, mulct as credo, should be critical for morality! But it is so. And all the more because of its conception, it having been born of the protest against the expanding doctrine of liberty, equality, and fraternity, the life blood of ethics. For the formulations of the

Page 212

b

theory by both Wallace and Darwin had the same source: <u>An Essay on the Principle of Population ...</u> (1798) by Thomas Robert Malthus inspired them—a book that came out of the travail of a revulsion against the too great danger of the spreading influence of the doctrines of the French Revolution. These and the earlier American Declaration of Independence had founded a new hope for human justice on the principle of the equality of man as earlier had Christ proclaimed the brotherhood of man. Thus, a foundation-theory of biology had the anomalous effect outside of biology of reversing a basic scientific postulate: that human equality is a biological truth.

Darwin later formulated a theory of human morality within the purview of evolution, one deriving morality from the social instinct as portrayed in the animal kingdom. This came too late to mitigate both the influence of his earlier writing in emphasizing the struggle for existence and even more, the view engendered by his followers in emphasizing this factor. As has been said by Fouillée, especially Darwin's twelve years of silence gave greater power to opponents of an extravagant interpretation of his theory by his followers. By a sort of poetical justice, Darwin's ethics was received in much the same silence which he maintained while his supporters exaggerated his theory of evolution. Once the human mind has accepted a theory, however much a tentative and approximate explanation is proffered, often the more so the more it is tentative and approximate, it is difficult to accomplish mental disgorgement. Our minds are slow to receive new ideas—Anatole France, with his genius for irony, makes one of his characters give the advice to another, a prospective writer of history, not to attempt to give new ideas. On the other side, most natural scientists will doubtless agree with Lamarck who said it is good that the mind

Page 213

c

does not too quickly embrace the new. Certainly, in the words of Bernard Shaw, it is dangerous to take on the new whilst retaining the old when they are in conflict. But once having accepted a new idea our minds are still slower in abandoning it. False notions frequently are more readily translated into belief than those nearer the truth. Not that the struggle for existence is as such false—to the contrary, it is a self-evident truth, a banal, an every-day commonplace. What is false is the putting of it in juxtaposition with another self-evident truth, the progressive grades of complexity in the living world, together with a hypothetical picking out agency. For, if you grant this all wise <u>type-setter</u>, then you may discard the struggle for existence.

The idea of evolution, having prevailed through centuries, it is not necessary to center the present exposition of man's ethics as a derivative of the evolutionary process on any definite date. Furthermore, this essay, making no pretense at being an historical treatment, need not follow a strictly chronological order, which would offer certain psychological hazards to thinking. My interest is in the <u>ideas</u> concerning ethics rather than in the history of ethics. That theories of man's ethical behavior are limited neither in time nor by national boundaries is an additional reason for discarding a historical method of treatment. In all times and in all countries of Western Europe, thinkers have postulated an ethics as separate from religion, at the same time either denying or affirming its divine origin. Thus, it is a false dichotomy, with one camp affirming that ethics is a gift from God to man directly and the opposing camp denying this special

Page 214

d

and divine gift. And as has been said above, men, including Wallace even, who were proponents of the theory of evolution, were loathe to accept the extension of this principle to include man's spirit. The strongest apostle of Darwinism in England, Thomas H. Huxley, in his famous Oxford lecture, entered himself as a subscriber to the belief in the divine origin of man's ethics.

An ethics esteemed as a blossom on the tree of knowledge ought perforce derive its sustenance, as the tree itself, in part from the roots, the natural sciences. It ought in no wise be an adventitious efflorescence, in the sense, that is, of an epiphenomenon or a chimera, a product of a graft from unknown source. Then a materialistic morality ought have ground-support as well as a portion of its nourishment from the natural sciences. These sciences include biology. But because of the amazing triumphs of physics, we tend to give physics hegemony to overlook biology, and chemistry even. Materialism in physics has thus come to mean materialism in all the natural sciences, especially on the part of many biologists who blindly accept the mechanistic doctrine of physics. The modern version of this materialism has been clearly exposed by Bertrand Russell:

> "That man is the product of causes which had no prevision of the end they are achieving; that his origin, his growth, his hopes and fears, his loves and beliefs, are but the outcome of accidental collocations of atoms; that no fire, no heroism, no intensity of thought and feeling, can preserve an individual life beyond the grave; that all the labours of the ages, all the devotion, all the inspiration, all the noonday brightness of human genius, are destined to extinction in the vast death of the solar system, and that the whole temple of Man's achievement must inevitably be buried beneath the debris of a universe in ruins— all these things, if not quite beyond dispute, are yet so

nearly certain that no philosophy which rejects them can hope to stand." [From *Why I am Not a Christian* (1927) by Bertrand Russell]

Page 215

e

But now it seems to appear that this materialism is no longer quick or is at least moribund. If it is truly dead, then a materialistic ethics sustained by the mechanistic concept is no longer wholly viable. Suppose, however, that this concept be true; can it exercise an inhibiting effect on the unfolding of human decency? I should say that the corroding power of this etching process of physics need not extend beyond the ground where its province is. There are other roots than physics. Despite the inept strivings of the mechanistic biology, the science of life remains apart from the science of non-life. Whether the mechanistic conception is quick or dead in physics, it is aborted in biology. There is no morality in the non-living world, its materialism or non-materialism has no direct channel to ethics. As a constant of the living state, co-existing with life itself, ethics can be materialistic only if we accept materialism as a ground-principle in biology. Hence, I hold that there can be no materialistic ethics.

Attracting adherents since the time of the Greeks, thoroughgoing materialism has retained a foothold, its tenets expressed in terms of knowledge won. Its formulation, as that by Bertrand Russell, is only more far-reaching because of the greater proof of its extensive practicability and its enormous success in modern times. By these criteria it has been regarded as a revelation of final truth by many others than Bertrand Russell. Indeed, lured by its triumphs in practical application, many outside the fields of science in this age of the radio and the aeroplane have taken it for its large cash value as the final word of eternal verity. Others, somewhat bewildered have sought

Page 216

f

refuge in the life of a curious sort of split personality. To them the world is dual: one, a world of the unalterable mechanism of physics; the other, transcendental, so to say, a revelry of sound where laughter need not exist, a riot of color where wave-lengths are unknown, and a flow of emotion where the tears of one sighing for lack of new worlds to conquer and of another weeping over creant Jerusalem are held to be something other than like chemical solutions. Yet again, others give themselves over to despair, seeing in this doctrine of the <u>metamechanicians,</u> who by so extreme abstraction of facts have in physics a realm comparable to metaphysics, an atmosphere exposed to which their dreams deliquesce. But the true man of science has little concern with human hopes and fears and less with the fragile stuff of dreams; to him scientific fact alone is fact. However much it destroys our dreams, this fact must prevail. Hence the dogmatism of Bertrand Russell.

Darwin sets forth his theory of the origin of ethics in chapter four of <u>The Descent of Man</u> which begins thus:

"I fully subscribe to the judgment of those writers who maintain that of all the differences between man and the lower animals, the moral sense or conscience is by far the most important. This sense, as Mackintosh remarks, "has a rightful supremacy over every other principle of human action;" it is summed up in that short but imperious word ought, so full of high significance. It is the most noble of all the attributes of man, leading him without a moment's hesitation to risk his life for that of a fellow-creature; or after due deliberation, impelled simply by the deep feeling of right or duty, so sacrifice it in some great cause. Immanual Kant exclaims, "Duty! Wondrous thought, that workest

neither by fond insinuation, flattery, nor by any
threat, but merely by holding up the naked law in

Page 217

g

the soul, and so extorting for thyself always reverence, if not always obedience; before whom all appetites are dumb, however secretly they rebel; whence thy original?"

This great question has been discussed by many writers of consummate ability; and my sole excuse for touching on it, is the impossibility of here passing it over; and because, as as I know, no one has approached it exclusively from the side of natural history. The investigation possesses, also, some independent interest, as an attempt to see how far the study of the lower animals throws light on one of the highest psychical faculties of man.

The following proposition seems to me in a high degree probable—namely, that any animal whatever, endowed with well-marked social instincts, the parental and filial affections being here included, would inevitably acquire a moral sense or conscience, as soon as its intellectual powers had become as well, or nearly as well developed, as in man."

Discussing sociability, Darwin points out first that even distinct species live together: some American monkeys; united flocks of rooks, jackdaws and starlings; man and dog. Passing over those insects that "are social and aid one another in many important ways", he next gives attention to higher social animals. These exhibit sociability by warning each other of danger "by means of the united sense of all"; perform many little services one for the other, licking the itching spot, removing external parasites, etc.; some predatory animals hunt in packs and aid each other in attacking their victims; they defend each other, as do the bull bison the cows and calves and the male baboons the troop, in times of danger.

Darwin next asserts that "it is certain that associated animals have a feeling of live for each other, which is not felt by non-social adult animals." But although he considers it more doubtful that they actually sympathize in the pains and pleasure of others, he tells us that they certainly sympathize with each other's distress or danger. To support this statement,

Page 218

h

he cites reported cases of birds aiding each other as that of an old and completely blind pelican, found by Captain Stansbury, "which was very fat, and must have been fed by his companions." He adds a personal observation of "a dog who never passed a cat who lay sick in a basket, and was a great friend of his, without giving her a few licks with his tongue, the surest sign of kind feeling in a dog."

So too, says Darwin [in *The Descent of Man, and Selection in Relation to Sex* (1871)], that which "leads a courageous dog to fly at anyone who strikes his master" must be called sympathy. A small monkey defending a keeper at the Zoological Garden against a fierce baboon is another instance of sympathetic and heroic conduct. "Besides love and sympathy, animals exhibit other qualities connected with the social instincts, which in us would be called moral" (Darwin 1871).

Dogs possess some powers of self-command, fidelity and obedience. The elephant also is faithful to his driver or keep "and probably considers him as leader of the herd." Next follows an observation, made by Dr. Hooker, which Darwin calls "a wonderful proof of noble fidelity" of an elephant to his keeper.

All animals that live in a body and defend themselves or attack in concert must be faithful to each other; if they have a leader they must be in some degree obedient. What leads animals to herd and to aid one another, we may infer, says Darwin, is the same sense of satisfaction that results from the performance of other instinctive actions. "It has often been assumed that animals were in the first place rendered social, and that they feel in consequence uncomfortable when

Page 219

i

separated from each other, and comfortable whilst together; but it is a more probable view that those sensations were first developed, in order that those animals which would profit by living in society, should be induced to live together, in the same manner as the sense of hunger and the pleasure of eating were, no doubt, first acquired in order to induce animals to eat. The feeling of pleasure from society is probably an extension of the parental or filial affections; … and this extension may be attributed in part to habit, but chiefly to natural selection. With those animals which benefited from living in close association, the individuals which took the greatest pleasure in society would best escape various dangers; whilst those that cared least for their comrades, and lived solitary, would perish in greater numbers. With respect to the origin of the parental and filial affections, which apparently lie at the base of the social instincts, we know not the steps by which they have been gained; but we may infer that it has been to a large extent through natural selection." To Darwin, "parental affection, or some feeling which replaces it, has been developed in certain animals extremely low in the scale, for example, in star-fishes and spiders."

Darwin now points out that the emotion of sympathy is distinct from that of love. Through natural selection sympathy increased; those communities in which there were the largest numbers of sympathetic members would flourish best and rear the greatest number of offspring. However, it is not possible in many cases to decide "whether certain social instincts have been acquired through natural selection"; they may be the indirect result of

Page 220

[letter j is skipped]
k

other instincts and faculties, such as sympathy, reason, imitation or the result of long-continued habit.

Turning to man, Darwin sets forth the idea that man as he now exists retains few special instincts, for he has "lost any which his early progenitors may have possessed;" yet he may "have retained from an extremely remote period some degree of instinctive love and sympathy for his fellows." Man as a social animal almost certainly would "inherit a tendency to be faithful to his comrades, and obedient to the leader of his tribe"—qualities common to most social animals." He would consequently possess some capacity for self-command. He would from an inherited tendency be willing to defend, in concert with others, his fellow-men; and would be ready to aid them in any way which did not too greatly interfere with his own welfare or his own strong desires."

"The social animals which stand at the bottom of the scale are guided almost exclusively, and those which stand higher in the scale are largely guided, by special instincts in the aid which they give to members of the same community; but they are likewise impelled by mutual love and sympathy, assisted apparently by some amount of reason." "The social instincts, which must have been acquired by man in a very rude state, and probably even by his early ape-like progenitors, still give the impulse to some of his best actions."
[Darwin (1871) *The Descent of Man, and Selection in Relation to Sex.*]

This work of Darwin on the origin of man's moral sense is so well known that the brief sketch of it just given suffices for the present purpose. To quote in their entirety the pages that Darwin devotes to this subject is thus unnecessary. Besides, I doubt if such inclusion would serve any the more to facilitate my

Page 221

1

analysis of his idea since what he has written includes matter not strictly germane to the points that I intend to expose. To offset any possible lodging of a charge of unfairness, I may say that I have endeavored to present Darwin's arguments faithfully and in the order which he followed.

What then was Darwin's conception of the origin of man's ethical behavior, what he called the moral sense?

In the first place, the co-founder of the theory of natural selection again and again invokes this principle as that which is at the bases of his idea as brought out in the summary of his argument sketched above. Moreover, in Chapter Five, continuing his exposition, he recurs to the role of natural selection in perfecting the intellectual and moral faculties of man and the social qualities of sympathy, fidelity and courage. Where in other passages the principle of natural selection is not set forth definitely in words, it is clearly implied.

On the other hand, there are certain passages that indicate or even state that natural selection may not be the sole operable factor. For example, we are told that certain social instincts may be the indirect result of other instincts and faculties or the result of habit. And again, in Chapter Five it is said that the bravest men willing to go to war and freely to risk their lives for others would perish in larger numbers than other men. "Therefore, it hardly seems probable, that the number of men gifted with such virtues, or that the standard of their excellence, could be increased through natural selection, that is, by the survival of the fittest." However, despite these admissions of doubt, though the operation of the principle of natural selection,

Page 222

m

Darwin maintains, man has gained his moral sense.

Secondly, Darwin maintains in one place that the moral sense is aboriginally derived from the social instincts. At other times again he speaks of instincts although as often he refers to sociability as a single persistent, deeply planted or ever-enduring instinct. This lack of clarity as to singularity may be dismissed. What cannot be so readily overlooked is his uncertainty concerning what he calls the parental and filial affections; these are included with the social instincts, are extensions of them, or apparently lie at the base of them. This is in serious contradiction since Darwin maintains that both the moral sense and the social instincts relate at first to the community. It is difficult to see how the affection of parents for their offspring or that of offspring for parents relate exclusively to the community and less to the species on which natural selection operates. The contradiction becomes more serious when we appreciate Darwin's conception of the meaning of the term, parental affection. Says Darwin: "Parental affection, or some feeling which replaces it has been developed in certain animals extremely low in the scale, for example, star-fishes and spiders." [1871] (Underlining is mine.) Even if we substitute for the word, replaces, some other, as, antedates, we still must enter an objection to this sentence.

Among star-fishes, members of the phylum generally placed midway between the first phylum, Protozoa, and the highest to which man belongs and among members of the phylum of arthropoeds, which includes spiders, insects, etc., and which most zoologists place next to the chordates, the highest species of which is man, occur some species that possess external structures, brood-

Page 223

n

pouches and the like, in which the young pass through their early stages of development. Several species of marine worms, as those of <u>Autolytus</u>, and of <u>Spirobis</u>, as well as serpent star-fishes on all of which I have made observations in order to learn their breeding habits, have such pouches. Similarly, the males of the pipe-fish and of its ally, the sea-horse, carry the eggs until they hatch. The so-called obstetric toad that carries its young furnishes another example of the external attachment of eggs to the parent. That the female kangaroo places her newborn young in her pouch or marsupium is well known. These examples are all taken from animals placed rather high upon the scale of life. In addition, may be cited cases of hydroids, members of the third phylum of the animal kingdom, from whose reproductive branches, or hydrants, the young (medusae or jelly-fish) are liberated. Moreover, many animals give birth to living young. But this viviparous mode is by no means confined to the higher forms; for example, in sponges, constituting the second animal phylum, the embryos develop in the parent's body; when they leave it, they lead a short free-swimming existence before they settle down as sessile organisms, from each of which arises the colonial sponge. Add finally instances of animals, as birds, that incubate their eggs; and those of fishes, as the common toadfish, which guard the eggs.

Surely no one of these types of behavior can be denominated parental affection or a feeling that replaces it, any more than intra-uterine gestation in mammals, can be attributed to the mother's love. There are, at least for humans, indubitable cases of gestation occurring against the wish and despite the hate of the mother. An external brood pouch in a serpent starfish, in

Page 224

o

a worm or in fishes, is a structure the use of which has nothing to do with parental affection, even if you grant, which I do not, that starfishes and spiders possess such or a feeling that replaces it. These are structural adaptations to be explained as such. And the explanation probably differs so greatly that a single explanation would scarcely hold for all cases. Thus, observations of my own made in 1913 on the pipe-fish indicate that the eggs in the brood undergo development in a medium of different osmotic pressure from that of the sea-water since, when removed from the pouch they develop only in sea-water having been diluted. This is not the case for eggs of worms. At any rate it is not paternal affection or maternal solicitude that induces the male to carry the eggs. It is not parental affection that hold the eggs in place on the abdomen of a lobster but a sticky secretion. A hen of the common barn-yard fowl will hatch out the eggs of a duck having in the meantime sat on a porcelain egg of neither of which she is the parent. Eggs of a sponge having been dissected out into sea-water develop as well as in the sponge's body. Instinct doubtless is responsible for the kangaroo's action in placing her new born young in the marsupium, an action that does not imply affection, however. There remains a still more powerful objection to Darwin's idea of parental affection, be it with, from or below the social instinct or instincts.

For every one of the examples given and for all possible examples that can be given of parental care are more to indicate that animals fail to care for their eggs or their young. In the vast majority of cases, animals are oviparous, not viviparous,

Page 225

p

and the eggs, once emitted, are left to chance. The young of most viviparous animals outside the class of mammals must from birth fend for themselves. The number of cases of neglect of the young outweigh those of care. Here I mean care in the sense of instinctive action without any implication of parental love or affection. And it must not be overlooked that some of the most non-social care for their young with extreme solicitude. It must be recalled, on the other hand, that some most social creatures destroy their mates to feed the young.

As to filial affection, a word suffices. The helplessness of a young creature dependent upon its parents ought not to be counted as a manifestation of affection. The relation is one of physiological dependence solely. One might just as well assert that the embryo in utero attaches itself to the mother through filial affection whereas the means of attachment is a well-known structural entity.

Said briefly: the relation of eggs or of young to parent shown to be a structural dependence cannot be confused with a spiritual attribute as affection or any feeling that replaces it or antedates it.

But let us put aside this argument on the affections for it displays too great weakness in its zoology to warrant extended discussion. Even so, there remain shortcomings in Darwin's conception of the social instinct. Two of these I mention.

The first is that in discussing sociability, he continually errs in treating social behavior within and without the species as if the two were the same. Whilst he ought confine himself entirely to social relations among members of the same species,

Page 226

q

he dwells too much on the living together of distinct species: of monkey and of birds, of man with dog or elephant, or dog and a cat, of monkey and man, etc. This argument drawn from domesticated animals is especially weak because one may by including the host of animals trained for the circus prove that all animals are sociable. More, Darwin directly tells us that a man or boy risking his life to save a drowning man, though a stranger, is impelled by the same instinctive motive which made the heroic little American monkey, formerly described, save his keeper, by attacking the great and dreaded baboon." In other words, Darwin makes no distinction between symbiosis, commensalism, the living together of distinct species for mutual benefit,—be these in nature or as the result of domestication or training—and intraspecific sociality, which is or ought to be his subject.

The second short-coming that I would mention is that Darwin is not at all clear as to where first appears the social instinct or those instincts which he considers aboriginal to man's moral sense. We are told that any animal endowed with well marked ones would inevitably <u>acquire</u> the moral sense when he approached or or equaled man in intelligence. Further, we learn, the social instincts must have been <u>acquired</u> by man in a very rude state or probably by his early ape-like ancestors. The acquisition of the moral sense or of the social instincts being uncertain, the evolution is doubtless even more so. And yet the essay is a contribution to evolution (or wants to be!).

Now comes the third and final point of my analysis. This is that Darwin registers complete failure in his attempt to answer the great question, "discussed by many writers of consummate ability",

Page 227

r

of what is the origin of man's moral sense?   Hence Darwin's conception is null and void.

As to the social animals: the data if these could be accepted in toto are insufficient. They do not by any means include even a respectable number of those species that display instinctive actions but concern in largest measure the higher vertebrates and, of these, for the most part some mammals. If in treating man's descent one approaches the problem of ethics "exclusively from the side of natural history", one ought to survey the whole field and not merely a limited region of it. The natural history of the animal kingdom embraces twelve phyla; an ethics based on examples, some doubtful, drawn from examples of the sociability of a fraction of a sub-phylum is an ethics derived from the natural history of this fraction only.

Moreover, both among mammals, as we are aware, and, as Darwin himself points out, among other vertebrates, as well as certain groups of invertebrates exhibiting instincts, some species are social and others not. To derive man's moral sense from the social instinct, or instincts, it would be necessary to adduce evidence to show clearly that man has evolved from a line of social animals. Darwin could not do this even considering mammals alone. Far from being able to trace man's descent, he makes, in Chapter Two, this most damaging admission:

> "In regard to bodily size or strength, we do not know whether man is descended from some small species, like the chimpanzee, or from one as powerful as the gorilla; and, therefore, we cannot say whether man has become larger and stronger, or smaller and weaker, than his ancestors. We should, however, bear in mind that an animal possessing great size, strength, and ferocity, and which, like the

gorilla, could defend itself from all enemies, would not perhaps have become social; and

Page 228

s

> this would most effectually have checked the acquirement of the higher mental qualities, such as sympathy and the love of his fellows. Hence it might have been an immense advantage to man to have sprung from some comparatively weak creature." [Darwin, Charles (1871). *The Descent of Man, and Selection in Relation to Sex.*]

In other words, the paucity of the examples given of social animals limited too greatly to the highest forms far from supporting his argument lays it open to serious objections. He renders himself guilty of the charge of loose thinking in this and in other ways, as even a cursory reading of the chapters here reviewed reveals. Then let us suppose that we know beyond question that man has descended from some social primate like the chimpanzee—this ancestor, please bear in mind, could however just as well have been a gorilla-like form since the gorilla, in caring for its dependent young, possesses what Darwin calls parental and filial affections, which he includes with, derives from, or bases upon, the social instincts, and is accordingly a social animal!—non-the-less the argument collapses. It does not give us evidence that in the progressive evolution in the animal kingdom, from the lowest form to man, the advancement was moved along the line of the social animals. No less than such evidence we demand of one who categorically states "that any animal whatever with well-marked social instincts, the parental and filial affections being here included, would inevitably acquire a moral sense." And it is no excuse to admit, as Darwin does, after having promised to "make some few remarks on the probable steps and means by which the several mental and moral faculties of man have been gradually evolved", that neither his ability nor knowledge permits the attempt "to trace the development of each separate faculty from

Page 229

t

the state in which it exists in the lower animals to that in which it exists in man." A defender of his might protest here, saying that these two passages ought not to be put in juxtaposition, for the first, concerning gradual evolution, refers to man's development, i.e., individual evolution. I am willing to admit this as a possible interpretation in view of the next sentence, which clearly refers to the developing infant and to the "perfect gradation from the mind of an utter idiot, lower than that of an animal low in the scale, to the mind of a Newton." I am further willing to overlook the phrase, "lower than that of an animal low on the scale", if the hypothetical defender so wishes. Then I rest on what to me is a fundamental proposition: an evolutionist—to many minds, the evolutionist—in an essay on the evolutionary descent of man exposing any phase of the evolution-problem ought in unequivocal terms exhibit unmistakably the locus of this phase and the congruity of this exhibition with the whole problem and with his conception of operable factor or factors. Instead, Darwin writing on the origin of man's ethical behavior, is an exhibition of dilettantism, weak enough on the side of natural history and made weaker now by being misplaced.

Darwin would have contributed a splendid chapter to the history of human thought had he succeeded in establishing his thesis that man's ethical conduct is the resultant of the evolutionary process. The destiny of humanity might have been altered had he been able to prove that our ethics is derived from the social instinct as portrayed in the brute-world. Despite his failure, he is due all merit for having attempted

Page 230

u

to formulate an ethics in terms of the concept of evolution. And if in the foregoing I have adversely criticized his argument—all to sharply, his admirers may say—I do not for a moment withhold admiration for his effort. When you estimate a man's work, you appraise the whole and in so doing strive to exhibit the good and the positive, which alone make for advance, rather than the error and the lack, thus giving less place to failures than to gains. No amount of dispraise can impugn Darwin's worldwide reputation. Nevertheless, it is often necessary to expose a faulty conception; most of all in science neither a theory nor its founder should be sacrosanct—far less should either be a national institution to be worshipped, especially when an idea becomes, as all good ideas become, the possession of mankind generally, its promulgator losing all except a claim for his individual method of presentation. This is what Buffon meant in his famous discourse on style when he said that style is the man himself; the idea at the basis of his scientific discovery becomes universal property. Darwin has said: "False facts are highly injurious to the progress of science, for they often endure long; but false views, if supported by some evidence, do little harm, for everyone takes a salutary pleasure in proving their falseness; and when this is done, one path toward error is closed and the road to truth is often at the same time opened."

Darwin's attempt to found an ethical theory on the evolution of social instinct was generally overlooked even, or especially, by his disciples. Indeed, Huxley, who was certainly the strongest proponent of Darwinism, denied his master's teachings on ethics

Page 231

v

for in his famous Oxford lecture delivered in 1893, twenty-two years after the appearance of <u>The</u> <u>Descent</u> <u>of</u> <u>Man</u>, Huxley said:

> "Social progress means a checking of the cosmic process at every step and the substitution for it of another, which may be called the ethical process; the end of which is not the survival of those who may happen to be the fittest, in respect of the whole of the conditions which obtain, but of those who are ethically best."
> [From *Evolution and Ethics* (1893) by Thomas Henry Huxley; Kindle edition p. 38]

> "Let us understand, once for all, that the ethical progress of society depends, not on imitating the cosmic process, still less on running away from it, but in combatting it." [Huxley 1893: 39]

When later published there was appended to the first of the sentences quoted the following note [Note 20]:

> "Of course, strictly speaking, social life, and the ethical process in virtue of which it advances toward perfection, are part and parcel of the general process of evolution." [Huxley 1893: 55]

But this note, presumably an afterthought on Huxley's or some other's part, could not recall either the passage noted or the entire lecture, which was strictly in keeping with Huxley's pronouncement made five years earlier in which he said, "speaking from the point of view of the moralist" that "the animal world is on about the same level as a gladiator's show. The creatures are fairly well treated, and set to fight; whereby the strongest, the swiftest, and the cunningest live to fight another day. The spectator has no need to turn his thumb down, as no quarter is given." [From "The Struggle for Existence: A Programme" (1888) by Thomas H. Huxley.]

And later, speaking of primitive man, he tells us that "the weakest and stupidest went to the wall, while the toughest and shrewdest, those who were best fitted to cope with their circumstances, but not the best in another way, survived. Life was a continuous free fight, and beyond the limited and temporary relations of the family, the Hobbesian war

Page 232

w

of each against all was the normal state of existence."

The reason why Darwin's conception of ethics as a culmination of the evolutionary process failed to obtain favorable reception is thus not far to seek: the picture of living nature, including primitive human society, as one interminable struggle, of bloody fight, of murderous and wholesale warfare, was too strongly engraved to be easily expunged. Huxley's repudiation of Darwin's ethical theory, for such it is despite the addendum on sociality, proves how strongly engraved was the doctrine of struggle in animate nature. For it was Huxley, more than anyone else, not excluding Haeckel, who was the great disseminator of Darwinism. His lecture on ethics had therefore the authoritative weight of an exegete, it implied a sort of reproof of the master guilty of nodding.

Other reasons might be proffered for the failure of Darwin's idea of the origin of man's ethical behavior. Since many who, though they subscribe to the principle that living things have evolved, deem man's spirit or soul outside the framework of organic evolution, they would reject this view of the derivation of ethics, preferring to regard the soul and its manifestations as divine gifts to man alone. The theory, interpreted as one separating ethics from religion, might be distasteful to those who deny the independence of ethics from religion. The suggestion that human morality is but an extension of the social instinct, as displayed in the brute-world, could be thought of by some as novel and perhaps far-fetched, lacking sufficient evidence, as I have indicated above. But these suggested reasons, I think, once examined can be rejected.

Page 233

X

Ethics as independent of religion was no new thought in Darwin's time and least of all in Great Britain, which had produced many eminent moralists, including several theologians who had postulated that morality is separate from religion. In this—and in other respects as the late William Keith Brooks, that great biological thinker, pointed out—Darwin owed much to the "natural theologian." One calls to mind the clergymen Cudworth and Cumberland, bishop of Peterborough, together with the non-clericals, Hobbes and other philosophers. Shaftesbury, strong opponent of Hobbes, as were Cudworth and Cumberland, belongs in this group. On the other side of the channel, in France, had been thinkers like Pierre Charron, an advocate who later took orders; in Holland, Hugo Grotius, the father of international law, and Spinoza, who, it has been said, was drunk with God;—these and a host of others from the time of the Greeks onward postulated a morality independent of religion. And, mark you, most of these men were deeply religious. Add the brilliant array of French moralists who catalyzed the French revolution and the no less imposing list of those of the early nineteenth century, ending with Auguste Comte and the flowering of positivism, and you have again others, among whom some were definitely atheistic, who insisted upon a morality independent of religion. Not that this point of view was ever unanimously held—far from it; for always there were school of thought that endeavored to amalgamate ethics and religion. Following the enunciation of the principle of the struggle for existence came a recrudescence of this thought: many sought in deep religious feeling a veil against the horrific aspect of this Medusa, the implied amoralism of the pitiless

Page 234

y

war that Huxley so graphically portrayed. But it is to the early religionists [Charron, Hugo Grotius, Cudworth, Cumberland, Spinoza] rather than to the atheistic thinkers that I invite attention.

Finally, the idea of deriving ethics from the social instinct had long before Darwin more than once been hinted. Examples include Shaftesbury, who was combatting Hobbes's view of man as wolf to man, and Kropotkin, who quotes Francis Bacon as follows:

> "All things are endued with an appetite for two kinds of good—the one as a thing is a whole in itself, the other as it is a part of some greater whole; and this latter is more worthy and more powerful than the other, as it tends to the conservation of a more ample form. The first may be called individual, or self-good, and the latter, good of communion. ... And thus it generally happens that the conservation of the more general form regulates the appetites."
> [Francis Bacon according to *Ethics: Origin and Development* (1924) by Peter Kropotkin]

And later:

> "... [There are two] appetites (instincts) of the creatures: (1) that of self-preservation and defense, and (2) that of multiplying and propagating, ... The latter, which is active, and seems <u>stronger</u> and more worthy than the former, which is passive."
> [Bacon according to Kropotkin (1924)]

Kropotkin interprets this passage to mean that Bacon estimated the social instinct as the basis of morality. However, the Austrian, Jodl, as Kropotkin himself points out, in his very complete history of ethics gives the passage another meaning. Thus this derivation of morality from the animal instinct for sociality was not a new thought; its novelty was not the reason for is failure to attract adherents (when proffered by Darwin). This conclusion is borne out by Kropotkin's inability to succeed where Darwin failed. Neither his <u>Mutual Aid: A Factor of Evolution (1902)</u>, nor his <u>Ethics: Origin and Development</u> (1924), despite their truly admirable qualities, can be said to have accomplished more than Darwin's effort to establish morality as the evolutionary resultant of

Page 235

z

cooperative sociability among animals.

I repeat therefore my thought, namely, that the vast sea of public approval that carried high the Leviathan, the struggle for existence, as launched by Huxley and other apostles of Darwinism, swamped the little craft, mutual aid. Or, the tide had been taken at the flood by the former, and the latter was put out to sea too late. The first came safely to the port that harbors all successful ventures in ideas; the second, for all its kindly burden, found no refuge.

Kropotkin made a valiant attempt to dislodge the popular interpretation of Darwin, an interpretation wrongly proffering that "struggle for existence" is the antinomy of "mutual aid."

In *Mutual Aid: A Factor of Evolution* (1902) Peter Alekseyevich Kropotkin begins by recalling two aspects of animal life that impressed him during his youthful journeys in Eastern Siberia and Northern Manchuria; one, the extreme severity of "the struggle for existence, which most species of animals conduct against an inclement nature," and the other, the absence of "that bitter struggle for the means of existence <u>among</u> <u>animals</u> <u>belonging</u> <u>to</u> <u>the</u> <u>same</u> <u>species</u>, which was considered by most Darwinists (though not always by Darwin himself) as the dominant struggle for life (underlining here and in subsequent quotations are Kropotkin's), and the main factor of evolution" [p. 1]. In the coming together on lakes of scores of species and millions of individuals to rear their progeny, in the colonies of rodents, in the migration of birds along the Ussuri and of fallow-deer on the Amur, he saw "Mutual Aid and Mutual Support carried on to an extent" which made him "suspect in it a feature of the greatest importance for the maintenance of life, the preservation of each species, and

Page 236

z1

its further evolution" [p. 2]. Further, says Kropotkin, his observations on semi-wild cattle and horses, ruminants, etc., showed that when animals have to struggle against scarcity of food (in consequence of the severity of the Eursian climate), they come "out of the ordeal so much impoverished in vigour and in health that no progressive evolution of the species can be based upon such periods of keen competition" [p. 2]. In consequence, Kropotkin later could not agree with the relation between Darwinism and Sociology set up by writers who endeavored to prove that although man may mitigate the harshness of the struggle for life between man, "the struggle for means of existence of every animal against all its congeners, and of every man against all other men, was a 'law of nature'" [p. 2]. To Kropotkin it was necessary in order to combat the idea of "harsh pitiless struggle for life" to indicate "the overwhelming importance which sociable habits play in Nature and in the progressive evolution of both the animal species and human beings" [p. 6].

Following these statements, from the introduction, are chapters on mutual aid among animals and among men during various stages of human development. Kropotkin says:

> "Sociability is as much a law of nature as mutual struggle. Of course it would be extremely difficult to estimate, however roughly, the relative numerical importance of both these series of facts. But if we resort to an indirect test, and ask Nature: "Who are the fittest: those who are continually at war with each other, or those who support one another?" we at once see that those animals which acquire habits of mutual aid are undoubtedly the fittest. They have more chances to survive, and they attain, in their respective classes, the highest development of intelligence and

bodily organization. If the numberless facts which can be brought forward to support this

Page 237

z2

> view are taken into account, we may safely say that mutual aid is as much a law of animal life as mutual struggle, but that, as a factor of evolution, it most probably has a far greater importance, inasmuch as it favours the development of such habits and characters as insure the maintenance and further development of the species, together with the greatest amount of welfare and enjoyment of life for the individual, with the least waste of energy." [Kropotkin 1902: 11-12]

The chapters on animals teem with examples to show the prevalence of mutual aid. Kropotkin says:

> "With many large divisions of the animal kingdom, mutual aid is the rule. Mutual aid is met with even amidst the lowest animals, and we must be prepared to learn some day, from the students of microscopical pond-life, facts of unconscious mutual support, even from the life of micro-organisms. Of course, our knowledge of the life of the invertebrates ... is extremely limited; and yet, even as regards the lower animals, we may glean a few facts of well-ascertained co-operation." [Kropotkin 1902: 14]

Examples are next cited to show such cases among the "lower animals", i.e., insects and crabs, members of the highest phylum of invertebrates, the arthropods, standing next below the chordates. The social life of ants and of bees is extensively exposed.

What follows is a survey of mutual help for "all possible purposes" among higher animals, the birds being taken up first with the admission that our knowledge is still imperfect; the conclusion reached is that the war of each against all is not the law of nature. "Mutual aid is as much a law as mutual struggle ..." [Kropotkin

1902: 26]. As to mammals, Kropotkin is struck with the overwhelming numerical predominance of social species over those few carnivores which do not associate. But even among carnivores, social habits are found. Thus, "... life in societies is no exception in the animal world; it is the rule, the law of Nature ..." [Kropotkin 1902: 39].

Page 238

z3

Emphasizing social association and intelligence as a social faculty, Kropotkin says:

> "Association is found in the animal world at all degrees of evolution: and, according to the grand idea of Herbert Spencer, so brilliantly developed in Perrier's <u>Colonies Animales</u>, colonies are at the very origin of evolution in the animal kingdom. But, in proportion as we ascent the scale of evolution, we see association growing more and more conscious." [Kropotkin 1902: 40]

> "... life in societies is the most powerful weapon in the struggle for life, taken in its widest sense ... sociability is the greatest advantage in the struggle for life. Those species which willingly or unwillingly abandon it are doomed to decay; while those animals which know best how to combine, have the greatest chances of survival and of further evolution, although they may be inferior to others in each of the faculties enumerated by Darwin and Wallace, save the intellectual faculty. ... intelligence is an eminently social faculty." [Kropotkin 1902: 42-43].

> "The highest vertebrates, and especially mankind, are the best proof of this assertion. As to the intellectual faculty, while every Darwinist will agree with Darwin that it is the most powerful arm in the struggle for life, and the most powerful factor of further evolution, he also will admit that intelligence is an eminently social faculty." [Kropotkin 1902: 43]

"Therefore we find, at the top of each class of animals, the ants, the parrots and the monkeys, all combining the greatest sociability with the highest development of intelligence. The fittest are thus the most sociable animals, and sociability appears as the chief factor of evolution ..." [Kropotkin 1902: 43]

Thus, Kropotkin does not deny the principle of the struggle for existence. To the contrary, he says "no naturalist will doubt that the idea of a struggle for life carried on through organic nature is the greatest generalization of our century. Life is struggle; and in that struggle the fittest survive." [Kropotkin 1902: 44] Here Kropotkin devotes some pages to show that whilst "there is within each species, a certain amount of real competition for food—at least, at certain periods" [Kropotkin 1902: 45], this competition has neither the proportion nor plays the rôle assigned it in evolution. So his conclusion is: The

Page 239

z4

practice of mutual aid "is what Nature teaches us; and that is what all those animals which have attained the highest position in their respective classes have done" [Kropotkin 1902:53]. The argument supporting mutual aid among men is of the same tone.

It would be difficult to find a reader who would not be touched by this honest human heart burdened with kindness and tolerance, this essentially good man who felt that the theory of mutual aid could lend a humanizing touch to nature. How one wishes that Kropotkin was right! In all these intervening years since, in 1907, I read <u>Mutual Aid</u> for the first time, the character of the man has continued to impress me as much as the strong desire to bring myself wholly to agree with his theory. But desires are fragile wings that often beat themselves against the rigid bitterness of fact. With sorrow we must admit that there is lacking sufficient evidence to sustain Kropotkin's thesis.

Grant as we must that the term, struggle for existence, has various meaning other than that given by Darwin, for the most part, some kind of struggle, as Kropotkin admits, remains. Express this remainder in its simplest terms; is it then true that sociality is a weapon in this struggle, that it runs through the entire animal kingdom form micro-organism to man? Can we present an evolutionary scheme on the basis of a gradually ascending perfection of sociality so that the highest members of each class are the most sociable? These questions, I think, must be answered in the negative. We cannot grade sociality as we can grade animal forms from low to high. A colonial animal, moreover, is not, in the sense that Kropotkin used the term, a social animal. In addition, whereas protozoa, the lowest of all animals, are

Page 240

z5

often colonial, and sponges, the lowest of multicellular animals, are colonial, we cannot conclude that the colony-form is at the basis of social aggregates. The colony-mode of organization appears intermittently among animal groups above the sponges, some members of the highest phylum, the chordates, being colonial.

Social aggregates as colonial, are exponents of animal life that are here present and there absent. That mutual aid, cooperation, and sociality are tremendous aids in life cannot be gainsaid; that the evidence is insufficient to warrant the hypothesis that sociality parallels the course of evolution cannot be denied.

Nevertheless, <u>Mutual Aid</u> is an important work although opinion is here divided. Prenant holds it to be sentimental whereas Beales hails it as "a co-operator's classic, one of the sacred texts of Marxism and socialism, a book that may yet help to make an epoch" [Prenant; Beales]. At the same time, however, Beales rejects the carrying over to human society the principle of struggle for existence as interpreted by Darwin's followers—a principle, as we have seen, inspired by the Malthusian doctrine on human population—he accepts the idea of mutual aid among animals as apt for human society.

Kropotkin himself, in his <u>Ethics: Origins and Development</u> (1924), also protests against the application of the distorted interpretation of Darwin's metaphorical usage of the expression to human life. In this, I think, he was right, even though Darwin allowed this interpretation to persist without voicing a protest. But taken in the sense of Kropotkin, the struggle for existence remains too strong, too widespread; man's ethics as an expression of mutual aid is not the antinomy of the struggle for existence, the less so the more

Page 241

z6

acute you evaluate this struggle. As intraspecific and internecine war, as competition to the point of extinction, struggle for existence is an untenable principle both as a factor in evolution and as basis for morality. Thus, there is large measure of truth in Kropotkin's theory of the origin of ethics—and I like to think that, had he lived, he would have more closely approximated truth: he would, perhaps, have come to see that sociality, though not a invariable concomitant of gradation in the animal kingdom, was yet an exponent of something which from lowest animal to man parallels the successive degrees of complexity that is the chief and striking feature of the world of living things. For this is the inescapable fact, the very foundation-stone of the science of life: all theories of how this evolution came to be aside, the fact of evolution remains.

Thus, I march with Kropotkin for a long part of the way. And where I move on alone, I want to believe that as I travel I have before me the beacon of his truly magnanimous personality. For here there was a great soul, an idealist who could on bended knees kiss his native soil after long exile, but who died rejected by those who boasted to put in practice his unselfish life-long devotion to equity, justice and good-will among men. His own knew him not.

For long years I have speculated on the fact that so many of the foremost French thinkers have written on Joan of Arc, or on the French Revolution. Voltaire's Joan of Arc and Lamartine's *History of the Girondist* are examples; neither was, strictly speaking, a historian. Anatole France, also not a historian, wrote on both. Outside of France, to be sure, are notable treatments in

Page 242

z7

these subjects—Carlyle's *French Revolution*, Schiller's Joan of Arc, Mark Twain's Joan of Arc and his exquisite article which you may recall appearing in a magazine—Harper's I think it was—at the time of her canonization, and one of Bernard Shaw's many masterpieces, Joan of Arc. But it is the French themselves who recur periodically to these subjects. Why, I have asked myself?

The first answer is that they are both ideal, even mystical, subjects that engender a trajectory of the human soul beyond the narrow confines of the knowable, that lift up our being before the radiant light of something that ever shall we seek and never quite attain, yet are perhaps attainable as these phenomena show. Also, both came to end through sheer human brutality—a nullification of the spiritual purity, our real being, by the ugly being that we fasten to ourselves. Joan put to death by country-men of William the Conqueror in the capital of the land where once he reigned, not far from the street that today bears his name, and in which he himself met death; suffering before her release from the cruelty of persons who, as Mark Twain said, had the spirituality of a hen, while a miscreant king, whom she had saved, let her be.

No less spiritual in its conception was the French Revolution, and no less brutal was its post-parturitional life. The idealism of the Voltaire's, the Encyclopedists and the host of the pre-revolutionary thinkers reached its majority in the form of the prisoner of St. Helena, who in the words of Lord Rosebery, vainly beat his wings against prison walls, wings that had hovered overlong above his land, blighting it in the immense shadow that they cast. In the while, a tragic weak figure, moving

Page 243

z8

as if in a play, was doomed to languish in the palace of Schönbrunn, no less a prisoner albeit one in gandy dress. For the French Revolution <u>was</u> a spiritual phenomenon.

Thus, these two, Joan and the birth of the Rights of Man—pictures of the eternal verities that the human soul strives to encompass, and that the lust for power determines ever to withhold. There is yet another and deeper meaning to these phenomena as I think of them. But that is another story, one out of place here.

Kropotkin's idea of ethics was a dream and if, as dreams do, it lacked verity, it was nonetheless a facet of the truth. He sought to establish a counter-revolution against those forces that ever will crucify, put one who hears voices to the flames, and prostitute liberty, equality and fraternity. For the Darwinians had enthroned the principles that all spiritual phenomena in history have striven to dislodge. For Kropotkin, I should like to pay a cock to Esculapius.

[The End.]

Page 244

[Box 125-9, folder 162 contains, among other documents, 4-pages "Concerning various problems. Letter to P.U.   10.XI.33" [10 November 1933] signed by "m."]

[This letter to P.U. from "m" is about "a biologist," who is not a geneticist, "not an atomist, not a bio-chemist," and "not a vitalist." Instead, said biologist is committed to "exact experiment," believes that life stands "above physics and chemistry," that life is more than gene-chromosome activity or bio-physical-chemical reactions, and that life is more than the sum of its bio-physical-chemical parts. And "m" concludes "I want him to write out what he has in mind about these things," and "I want to help him in his experiments in which work he is alone ..." Perhaps "m" is for "muse," or for "Margret," or for "Maid." Margret Boveri says Hedwig Schnetzler introduced herself to E. E. Just as "Maid Hedwig Schnetzler" (Manning 1983: 204). Most likely, this letter was written by Maid Hedwig Schnetzler.]

-1-

Concerning various problems.
        Letter to P.U. 10.XI.33

None of the biologists knows yet what life is. To find the answer to this question is the goal of each workers investigation. But the way of study is very different. Also the conception they have of what may be the answer. Some see it in the direction of the little god "gene", they try to prove that we can reduce the life-phenomena onto a very small thing, that like an atom is the smallest "thing" we can find and in which the root, the primary cause of life is seated. Others put more interest in the changes that are brought about in the way of chemical reactions. They are trying to find a certain order of succession of such reactions in which they hope to find finally the primary reaction which sets on all the others in an ordered succession. The wonder that lies in the happenings and laws of chemical reactions they find and admire also in the living cell, which they conceive as a specially complicated and therefore alive chemical thing.

These two kinds of biologists (I do not mention other ones, since these two seem to me the chief ones) are determined to find a primary reason, either reaction or "life-atom", to which they might ascribe life as such. Quite consistent with the fact, that we can only, also in biology, work with physical and chemical, that is not-living, mechanical means, these workers say that all we can hope to find will be either a thing of physics or of chemistry. Nobody can tell yet whether they go in a wrong or in a right direction.

In contrast to these people stand the vitalists, who are much in the disadvantage because they have no experimental evidence for their speculations. Yet, we do not say that they are altogether stupid. Since the whole problem is still veiled, nobody has the right to reject any, even a speculative idea about it. And further: We all see that there is an essential difference between a living thing like an egg and a non-living thing like a ball. We therefore may doubt that even the finest analysis of the living thing with means of physics and chemistry will finally bring the answer to our question. It may very well be (as P. Umlein has pointed out in one of his letters) that such an analysis into the most minute parts of the cell as physical atoms or chemically reacting compounds may tell us at the end of a long way what we knew already at its beginning: namely, that life manifests itself <u>at least to us</u> in physicochemical forms of appearance.

Page 245

-2-   Letter   to   P.U.

10.XI.33

Some two years ago I met a biologist, who seemed to belong to none of the above mentioned groups. He believes that the evidence for a great importance of the chromosomes in the life-process is beyond doubt. But he will not ascribe omnipotence to them, nor to the genes. He does know very much chemistry (as I was often surprised to notice). But he is far from believing that by knowing the chemical processes in the cell we know what is life. Is he then a vitalist? Well, he has speculative ideas, his mind goes beyond the limits of what we have in certain knowledge. But never did I hear him talk of a "principle of life" or a Lebenskraft or any such things. On the contrary, nothing seems to be so close to his heart than the exactness of the experiments. Much of the classic work would he like to repeat out of doubt, that it might not come up to our postulation of exact experiment. (Never does he, for example, base a whole theory on some few experiments with one or two eggs.) So, he is not a vitalist either.

Some notion of this man's character made me understand better his peculiar point of view about his work and about the direction, in which the answer to the question about life may lie. This worker is indeed more modest than it is usually the case even with a very sympathetic worker of our times of the great scientific successes. He reminds me more of a worker of older times, when people were not yet drunk by what they could do with their mind and their hands but always had high appreciation for Nature which they thought and felt to be greater than they, the investigators. Closely connected with this modesty, but rooted even deeper in his soul is my workers love for Nature, for tiny animals in special. Many workers think most of their success. They love their work if it brings them fame and money. Others, the better ones, love it as an arena of the power of their mind, they put the glory of reason and man higher than the thing they work on. The biologist I am talking about has neither ambition nor does he think so much of his own mind. A dividing egg is to him a greater wonder than a dignified professor whose mind is so perfect.

Without this peculiar and great modesty and love, this man might have developed into a geneticist, a bio-chemist or a vitalist. Now, as he is, his modesty keeps him from expecting too much of mere science, of physics or chemistry. He does not think that we know even in these exact sciences already so very much that would warrant too great expectations. And of biology he thinks that it stands just at the beginning of its way. So he does not look for the "primary reason" of life, whilst we are

Page 246

still in the dark about so many second and third reasons. He thinks we should try to answer questions like: what is stimulation? Specificity? Fertilization? Before we look for a "life-atom" or a primary reaction. So my worker studies the cell-life in its minute steps and tries to follow it very exactly. Since he has so much love for these little wonders, his longing to find the answer to the big wonder that stands behind these is not so great. The geneticists seem to him to start at some speculative end-point before they have explored the country between their knowledge and this point. He wants to explore this country step by step. May be he can move only very slowly, because he does not know exactly in what direction he must move; everything is of interest and must be studied exactly. May be those who started at an assumed end-point, are finally right. Then his work, my biologists work, will nevertheless be of value. Because he has followed Nature on its natural path, which most probably we will be able to know only by such study, not by finding the "reason" for life. Because no matter how much we may know about this reason, life in its process will to our poor mind always be so complex, that only a study of every of its minute stages will help us to understand it.

Which is now that country, that my man does explore? It is the cell, the wonder of the developing egg, which he admires every day, if only he has a microscope and can marry sea-urchins. Quite in consequence with his modesty, that keeps him from looking for an end when he is at the beginning of his way, and with his love, that makes him admire an egg more than physics or chemistry, in which he feels still too much of our own mind and reasoning, he looks at the cell as a unit and as such standing aloof of the material it is made of – which material is furnished by physics and chemistry. The cell to him is not only a unit of structure, but also an entity because of the order of events that take place in it. Life manifests itself – so he believes – in the cell with regard to space in the peculiar way the cell is built – its peculiar structure, and with regard to time in the peculiar way this arrangement of its structure is changed. In bringing the structure of the cell and its changes together we may speak of the <u>movement in the cell</u>. This is, it seems to me, the most general expression for what he studies. We see, he is really not an anti-geneticist; the chromosomes, as minute parts of the cell which do express the cell's movement sometimes very clearly, do interest him. He appreciates the chemical reactions in the cell just as well. But to him,

Page 247

-4-

P.U. 10.XI.33   Letter to

as I have said above, these things are only parts of what interests him in his study of the entity of life, the cell. No other part in this entity must be neglected, not the interior of it, not the cortex. Of each of the factors he wants to know its proper place and movement in the cell. So he hopes to understand a bit better the physico-chemical manifestation of Nature as living cell.

He is not an atomist, not a bio-chemist, not a vitalist. What he is trying to understand with his reason – because he has it already in his fine instinctive feeling – is life as an entity, which stands above physics and chemistry, which it uses only to make itself seen to us. If therefore in our study we take the cell apart and study the atoms, or other stones out of which it is built, we have destroyed life and are physicists or chemists. Well, let us be such in order to find out the single stones out of which the great building of minute size is built, but let us not forget to study the whole building too, which is something more than a mass of stones. (See also letters of P. Umlein.)

Connected with this conception of life as a real entity (not simply a ragout of single things), as a Ganzheit, is my biologist's idea of the beginning of life in the moment, that a certain entity was formed by chance or otherwise in the mass of elements and molecules.

The fact, that my man's point of view is based on personal modesty and love as well as on some nice and exact experiments, does attract me. This is the reason why I want him to write out what he has in mind about these things, that is it why I want to help him in his experiments, in which work he is alone and which seems to me worthwhile to be extended.

In this sense let me close my considerations.

m. [hand-written]

Page 248

[Box 125-21, folder 396 contains, among other documents, these 3 pages.]

Paris, March 39.

-1-

The problem to be considered is the origin of ethics. The consideration treated is based on the postulate of evolution. Others have argued that ethics has evolved; here this evolution is carried farther back than heretofore has been done, and the relationship of this evolution with that of form and function is more strongly established.

We consider sex the result of a development on the principle of the division of labor. As special cell-regions developed that function in one exaggerated respect, so sex was formed and the labor in the process of perpetuation of the species divided. This separation entails: (a) the necessity of union of two individuals for the perpetuation of the kind; (b) the elongation away from the union of the many individuals who live in social communion, in other words, the "family" replaces the communion. The ethical respect, (a) develops love for children, family etc., (b) reduces social instincts for the benefit of individual ones. But why does the love of the mother for the child, the love of the members of the family for each other, finally lead back to sociability? In the case of the higher animals, reason certainly played its part in this development. In the case of lower animals (and in part this would hold also for the higher ones) the social instinct being older and more primitive than that developed through the division of labor by sexuality, is the force behind the newer form of relationship between the living beings. It thus turns the development of individual love back into sociability, or, otherwise said, prevents a branching off from the social tree of one, the extreme individualism, a-social instinct which would lead to complete isolation.

Page 249

-2-

Special Note: If behind the sexual love has thus always acted the older instinct—social instinct—then we may gain a different picture of the role of "natural selection." It was perhaps not so much the "fittest" who was picked out, not so much the "strongest," but the most sociable, the "best" with respect to sociability.

Behind love of an individual for another individual acts the social instinct. This instinct in its oldest form we call "interaction with environment." This environment is both inorganic and organic, and not living and living. It must be borne in mind that in this oldest form, this instinct cannot be called "love" or "social instinct." It is interaction and as such implies also repelling, rejecting, negating. Note here, that in the earliest stage what later appears as love and hate are simply directions of reactions: Repelling (hate) - Responding (love). Result: Maintenance of the species in cooperation with surroundings. The social instinct has its root in this interaction, which is the cooperation with the surroundings for the purpose of maintaining individual or species. The social instinct goes back to the necessity of cooperation between cell and surroundings. They are interdependent. If this interdependence is interfered with – by change of surroundings so that the cell cannot adapt itself any more, or by the change of the cell so that it loses its contact with the powers in the surroundings – pathology, abnormality or even death are the result. In other words, if the "repelling" goes too far, if hate wins over love, abnormality, pathology or death result.

Page 250

-3-

Social instinct therefore seems to be the instinct for the necessity of an interaction between individual and environment, an interdependence of the two, which must not lead to the exaggeration of the repelling factor. The reaction, the repelling, always goes only as far as the interdependence, the interaction demands or allows. Individual and surroundings are one interdependent unit, the social instinct is the instinct for this condition.

What we call "love" in the social instinct, in its oldest form is "response." The love of an individual for another one carries in it the old "response" of individual for surroundings. From here can be explained that unselfish factor in all love, and this explanation of unselfishness in love is better than that on the basis of mother for child, which after all can only hold for mothers, or at least only for girls.

Page 251

[Box 125-21, folder 396 contains, among other documents, this page.]

Roscoff,

28.IV.39

What is the <u>original</u> in our idea?

Surely, the evolution of the nervous system and precursors has been studied, and ethical behavior has been related to the nervous system. New in our idea is that a) we trace this evolution of the nervous system (and with it that of the ethical behavior) farther back than anybody heretofore. In so doing we complete the chain; b) we relate the endowments of the first link of the chain with those of the later and latest: the ectoplasm, and in so doing throw a new light on the meaning and character of ethical behavior. Man must act as comrade and fellow, altruistically, because he cannot help it. The roots of this feeling for fellow-man are older than family-love, older than sex-love. Ethical behavior of man reaches with its roots back to the very fundamental endowments of living substance.

We do not say that there is no struggle for existence. If there were no antagonism, there would never have been any evolution, there would be no ethics if man were not constantly confronted with having to choose between antagonistic feelings. There is endless struggle for existence in all life, also for the lilies. But there is little fight between species, less inside a species. The role of this kind of struggle has been exaggerated, just as that for the female in natural selection.

We take our support for the thesis of interdependence from the ectoplasm in evolution and in embryogenesis, and only later and in a lesser degree from the fact that sociable animals though the individual is week, do survive successfully (Kropotkin).

We do not want to discuss the origin of species; what interests us more is the origin of Genus and the evolution from lower to higher. (Darwin was particularly interested in the origin of species because of natural selection. Actually, what he means is the perpetuation of variations.)

[End of final page numbered 251.]

www.ingramcontent.com/pod-product-compliance
Lightning Source LLC
Chambersburg PA
CBHW030427160726
47991CB00005B/1625